Research at NaMLab

Band 4

Research at NaMLab

Band 4

Herausgeber:

Prof. Dr.-Ing. Thomas Mikolajick

Ekaterina Yurchuk

Electrical Characterisation of Ferroelectric Field Effect Transistors based on Ferroelectric HfO$_2$ Thin Films

Logos Verlag Berlin

λογος

Research at NaMLab

Herausgegeben von
NaMLab gGmbH
Nöthnitzer Str. 64
D-01187 Dresden

Bibliografische Information der Deutschen Nationalbibliothek

Die Deutsche Nationalbibliothek verzeichnet diese Publikation in der
Deutschen Nationalbibliografie; detaillierte bibliografische Daten sind
im Internet über http://dnb.d-nb.de abrufbar.

ISBN 978-3-8325-4003-6
ISSN 2191-7167

Logos Verlag Berlin GmbH
Comeniushof, Gubener Str. 47,
10243 Berlin

Tel.:   +49 (0)30 / 42 85 10 90
Fax: +49 (0)30 / 42 85 10 92
http://www.logos-verlag.de

Technische Universität Dresden

Electrical Characterisation of Ferroelectric Field Effect Transistors based on Ferroelectric HfO$_2$ Thin Films

Dipl.-Ing. Ekaterina Yurchuk

von der Fakultät Elektrotechnik und Informationstechnik der
Technischen Universität Dresden

zur Erlangung des akademischen Grades

Doktoringenieur
(Dr.-Ing.)

genehmigte Dissertation

Vorsitzender:	Prof. Dr.-Ing. habil. R. Schüffny	
Gutachter:	Prof. Dr.-Ing. T. Mikolajick	Tag der Einreichung: 23.07.2014
	Prof. Dr. K. Dörr	Tag der Verteidigung: 06.02.2015

Abstract

The ferroelectric field effect transistors (FeFETs) are considered as promising candidates for future non-volatile memory applications due to their attractive features, such as non-volatile data storage, program/erase times in the range of nanoseconds, low operation voltages, almost unlimited endurance, non-destructive read-out and a compact one-transistor cell structure without any additional access device needed. Despite the efforts of many research groups an industrial implementation of the FeFET concept is still missing. The main obstacles originate from the conventional perovskite ferroelectric materials (lead zirconium titanate (PZT) and strontium bismuth tantalate (SBT)), in particular their integration and scaling issues. The recently discovered ferroelectric behaviour of HfO_2-based dielectrics yields the potential to overcome these limitations. The decisive advantages of these materials are their full compatibility with the standard CMOS process and improved scaling potential. Utilisation of the $Si:HfO_2$ ferroelectric thin films allows to fabricate FeFETs in a state-of-the-art CMOS technology node of 28 nm. The ferroelectricity in HfO_2 has been discovered only several years ago. Therefore, there are still a lot of uncertainties about the origin of the ferroelectric behaviour as well as the impact of different fabrication conditions on its emergence. Moreover, the electrical behaviour of both the HfO_2-based ferroelectric films and memory devices based on these films requires more detailed studies. The emphasis of this work lays on the ferroelectric properties of HfO_2 thin films doped with silicon ($Si:HfO_2$). The potential and possible limitations of this material with the respect to the application in non-volatile FeFET-type memories were extensively examined.

The material aspects of the Si-doped HfO_2 thin films were studied at first in order to gain better insight into the occurrence of ferroelectricity in this system and to acquire guidelines for FeFET fabrication. The influence of the different process parameters such as the Si doping concentration, post-metallisation annealing conditions and film thickness on the stabilisation of the ferroelectric properties in $Si:HfO_2$ films has been examined. Electrical characterisation combined with structural analyses enabled the correlation of the changes in the macroscopic electrical properties to alterations in the film crystalline structure. The film composition was shown to have a strong impact on the electrical properties of $Si:HfO_2$ films. By varying the silicon doping level, paraelectric, ferroelectric or antiferroelectric-like behaviour was induced. Moreover, the temperature of the post-metallisation annealing as well as the film thickness can be used to tune the ferroelectric properties of the $Si:HfO_2$ films by changing their phase composition.

Furthermore, the electrical behaviour of the ferroelectric Si:HfO$_2$ films was analysed in detail. The effect of field cycling, polarisation switching kinetics and ferroelectric specific degradation (fatigue) were investigated. The improvement of the ferroelectric properties upon field cycling ("wake up" effect) that is often observed in perovskite ferroelectrics was also detected for Si:HfO$_2$ ferroelectric films. The polarisation switching times in the nanosecond range were ascertained. Fatigue properties of Si:HfO$_2$ films were shown to depend on the frequency and voltage amplitude. In contrast to perovskite ferroelectrics a dielectric breakdown was identified as one of the factors that limited the cycling capability of the Si:HfO$_2$ ferroelectrics. Due to operation at MHz frequencies and electric fields below 3 MV/cm the cycling capability was extended to 10^9 cycles and a fatigue-free behaviour was demonstrated up to 10^6 cycles.

The performance of the Si:HfO$_2$-based MFIS-FET devices, which were fabricated using the state-of-the-art 28 nm high-k metal gate CMOS technology, was investigated including the key memory characteristics, such as the program and erase behaviour, retention and endurance. The studied FeFETs demonstrated program and erase times in the nanosecond regime (10 – 100 ns) with operation voltages of 4 – 6 V. The operation capability of the Si:HfO$_2$-based ferroelectric transistors was proven in the temperature range between 25 and 210 °C. The retention behaviour of the studied devices deteriorated with increasing temperature and improved at higher operation voltages. Furthermore the impact of scaling on the memory performance of Si:HfO$_2$-based MFIS-FETs down to the gate length of 28 nm was investigated. The scaled devices demonstrated memory characteristics comparable to that of the long channel structures. The transistor short channel effect rather than deteriorated ferroelectric properties explained the observed difference in the behaviour of the scaled devices in comparison to the long channel devices.

The endurance (limited to 10^4 – 10^5 program/erase cycles) and charge trapping that is superimposed with the ferroelectric switching were identified as the main issues of the Si:HfO$_2$-based FeFET devices. A detailed study of both these issues was performed in this work. The limited endurance was found to be linked to the reliability of the transistor gate stack. A predominant degradation of the interfacial layer, which is embedded between the silicon substrate and the ferroelectric film, was detected. This was similar to the behaviour of the standard high-k metal gate stacks. Indications that this process can be held responsible for the endurance behaviour of the Si:HfO$_2$-based MFIS-FET devices were discussed. The electron trapping enhanced by the ferroelectric polarisation charge was shown to superimpose with the ferroelectric switching at typical erase operation conditions. This electron trapping impaired a fast erase. It is also suggested as the main cause of the interfacial layer degradation upon endurance cycling. A modified approach for the erase operation was proposed in this work in order to mitigate the impact of trapping and increase the effective erase speed.

Kurzzusammenfassung

Die ferroelektrischen Feldeffekttransistoren (FeFETs) zählen mit zu den vielversprechenden Kandidaten für zukünftige nichtflüchtige Speicheranwendungen. Der Grund dafür sind ihre attraktiven elektrischen Eigenschaften: die nichtflüchtige Datenspeicherung, Programmier- und Löschzeiten im Bereich von wenigen Nanosekunden, niedrige Betriebsspannungen, nahezu unbegrenzte Zyklenfestigkeit, nichtzerstörender Lesevorgang und eine kompakte Bauform der Speicherzelle, welche nur aus einem Transistor besteht und keine zusätzliche Zugriffsbauelemente benötigt. Trotz der Bemühungen von vielen Forschungsgruppen, konnte eine industrielle Umsetzung des FeFET-Konzeptes bisher nicht erreicht werden. Die Integrationsschwierigkeiten und Skalierungslimitierungen, die mit den klassischen Perowskit-Ferroelektrika (Blei-Zirkonium-Titanat (PZT) und Strontium-Wismut-Tantalat (SBT)) verbunden sind, sind dafür verantwortlich. Die vor kurzem nachgewiesene Ferroelektrizität von HfO_2-basierten Dielektrika bietet das Potenzial diese Einschränkungen zu überwinden. Der entscheidende Vorteil dieser neuen ferroelektrischen Materialien ist ihre Kompatibilität mit dem CMOS-Prozess in Kombination mit einem besseren Skalierungspotenzial. Die Anwendung von ferroelektrischen $Si:HfO_2$ Dünnschichten ermöglichte die Herstellung von FeFET Bauelementen auf einem aktuellen CMOS Technologieknoten von 28 nm. Die Entdeckung der ferroelektrischen Eigenschaften in HfO_2 Dünnschichten erfolgte erst vor wenigen Jahren. Infolgedessen gibt es immer noch sehr viele Unklarheiten über den genauen Ursprung dieser Eigenschaften und den Einfluss der verschiedenen Herstellungsprozessfaktoren auf ihrer Entstehung. Darüber hinaus muss ein besseres Verständnis vom elektrischen Verhalten, sowohl der HfO_2 ferroelektrischen Dünnschichten, als auch der auf ihnen basierenden Speicherelemente erworben werden. Das elektrische Verhalten von HfO_2 Dünnschichten, deren ferroelektrischen Eigenschaften mit Siliziumdotierung induziert wurden, stellte den Schwerpunkt der vorliegenden Arbeit dar. Das Potenzial und die möglichen Einschränkungen dieses Materialsystems, mit Bezug auf die Anwendung in nichtflüchtigen FeFET Speicherelementen, wurden ausführlich untersucht.

Als Erstes wurden die Materialaspekte der mit Si dotierten HfO_2 Dünnschichten mit dem Ziel analysiert, ein besseres Verständnis für das Auftreten von Ferroelektrizität in diesem System zu gewinnen. Die erworbenen Erkenntnisse dienten später als Leitpfaden für die Herstellung von FeFET Bauelementen. Der Einfluss von verschiedenen Herstellprozessparametern auf die Stabilisierung der ferroelektrischen Eigenschaften in $Si:HfO_2$ wurde untersucht. Unter anderem wurde die Konzentration von der

Siliziumdotierung, die Bedingungen für die Postmetallisierungstemperung und die Schichtdicke betrachtet. Die Kombination der elektrischen Charakterisierung mit den Strukturanalysen, ermöglichte die Änderungen in den makroskopischen elektrischen Eigenschaften mit den Änderungen in der Kristallstruktur zu korrelieren. Eine starke Abhängigkeit der elektrischen Schichteigenschaften von deren Zusammensetzung wurde beobachtet. Paraelektrisches, ferroelektrisches und antiferroelektrishes Verhalten konnte durch die Variierung der Siliziumdotierung induziert werden. Darüber hinaus wurden die Temperatur der Postmetallisierungstemperung und die Schichtdicke als zusätzliche Faktoren, die die Phasenzusammensetzung der $Si:HfO_2$ Dünnschichten und als Folge ihre ferroelektrischen Eigenschaften beeinflussen können, identifiziert.

Im nächsten Schritt wurde eine detaillierte Untersuchung der elektrischen Eigenschaften von den ferroelektrischen $Si:HfO_2$ Dünnschichten durchgeführt. Der Effekt des Feldzyklens, die Kinetik des Polarisationsschaltvorgangs und die Degradation, verursacht durch ein kontinuierliches Polarisationsschalten (Fatigue), wurden dabei analysiert. Ein elektrisches Wechselfeld verursachte eine Verbesserung der ferroelektrischen Eigenschaften von $Si:HfO_2$ ferroelektrischen Schichten. Dieser sogenannte "wake up"-Effekt ist auch charakteristisch für die klassischen Ferroelektrika. Polarisationsschaltzeiten im Nanosekundenbereich konnten für die untersuchten $Si:HfO_2$ Schichten nachgewiesen werden. Ihre Fatigue-Eigenschaften waren sehr stark von der Testfrequenz und der Amplitude des Stresssignals abhängig. Im Unterschied zu den klassischen Perowskit-Ferroelektrika, deren Zyklenfestigkeit hauptsächlich durch die Abnahme der schaltbaren Polarisation beeinträchtigt wird, hat sich für die $Si:HfO_2$ ferroelektrischen Schichten der dielektrische Durchbruch als Hauptlimitierungsfaktor erwiesen. Durch die Anwendung von Frequenzen im MHz-Bereich und elektrischen Feldern unter 3 MV/cm konnte die Zyklenfestigkeit von $Si:HfO_2$ verbessert werden, so dass 10^9 Schaltzyklen realisiert werden konnten. Eine nur geringfügige Degradation bei bis zu 10^6 Zyklen wurde festgestellt.

Die auf $Si:HfO_2$ Dünnschichten basierenden FeFET Speicherelemente wurden auf der Basis eines High-*k*-Metall-Gate-CMOS Prozesses im 28 nm Technologieknoten hergestellt. Ihre wichtigsten Betriebseigenschaften einschließlich der Datenhaltung, der Zyklenfestigkeit, des Programmier- und Löschverhaltens wurden untersucht. Die Programmier- und Löschzeiten im Nanosekunden-Zeitregime (10 – 100 ns) mit Betriebsspannungen von 4 – 6 V konnten nachgewiesen werden. Darüber hinaus wurde die Funktionsfähigkeit dieser Speicherelemente in einem Temperaturbereich von 25 bis 210 °C getestet. Mit zunehmender Temperatur wurde eine Verschlechterung der Datenhaltungseigenschaften festgestellt. Die höheren Betriebsspannungen hatten einen entgegengesetzten Effekt und bewirkten eine Verbesserung der Datenhaltung. Zusätzlich wurde der Einfluss der Skalierung auf die Funktionalität der auf $Si:HfO_2$-basierenden FeFETs analysiert. Zu diesem Zweck wurden die

Speicherelemente mit Gatelängen bis zu 28 nm hergestellt. Die hochskalierten ferroelektrischen Transistoren zeigten vergleichbare Charakteristiken zu den Langkanaltransistoren. Die Unterschiede in dem Verhalten zwischen Speicherelementen mit kurzen und langen Kanälen konnten größtenteils durch Transistorkurzkanaleffekte erklärt werden. Im Gegensatz dazu waren die ferroelektrischen Eigenschaften von der Skalierung nur geringfügig beeinflusst.

Die Hauptprobleme von auf Si:HfO$_2$ basierenden FeFET Speicherelementen sind zum heutigen Zeitpunkt die Zyklenfestigkeit (begrenzt auf 10^4 – 10^5 Programmier- und Löschzyklen) und der Ladungseinfang, der sich mit dem ferroelektrischen Schalten überlagert. In der vorliegenden Arbeit wurde eine ausführliche Analyse beider Aspekte durchgeführt. Es wurde festgestellt, dass Eigenschaften des Transistorgatestapels hauptsächlich für die limitierte Zyklenfestigkeit verantwortlich sind. Eine Ähnlichkeit zwischen den Degradationsvorgängen im ferroelektrischen und dem hoch-ε Gatestapel wurde festgestellt. Die Degradation während des kontinuierlichen Programmierens und Löschens beschränkte sich auf die Grenzschicht zwischen Substrat und hoch-ε Dielektrikum. Die Verschlechterung der Speichereigenschaften in den HfO$_2$-basierten FeFETs beim Programmieren und Löschen lässt sich durch die Gatestapeldegradation erklären. Ein weiterer Einflussfaktor ist der Elektroneneinfang während des Löschens, der durch die ferroelektrische Polarisation des Gatestapels zusätzlich verstärkt wird. Das hat zur Folge, dass sich die Elektroneninjektion vom Halbleitersubstrat bei typischen Löschbedingungen mit dem ferroelektrischen Schalten überlagert. Das gleichzeitige Einfangen von Elektronen beim Schalten, wirkt sich nachteilig auf die Löschgeschwindigkeit aus, die sich dramatisch reduziert. Das Einfangen von Elektronen wurde auch für die Degradation des Transistorstapels während des Zyklens und somit für die begrenzte Zyklenfestigkeit der ferroelektrischen Transistoren verantwortlich gemacht. Ein modifizierter Ansatz für die Löschoperation mit dem Ziel den schädlichen Effekt des Elektroneinfangs zu reduzieren und die effektive Löschgeschwindigkeit zu verbessern wurde in dieser Arbeit vorgeschlagen.

Contents

1 Introduction

Advancement in the field of microelectronics in the last several decades has revolutionised your daily life, where numerous electronic devices have become its inextricable part. The recent trend is a rapid popularisation of portable electronic devices (smart phones, digital audio players and digital cameras) and portable storage media (memory cards, USB flash drives, solid-state drives). A key enabling technology for this trend is non-volatile semiconductor memories (NVSMs) due to their capability of non-volatile data storage, speed, compactness, mechanical robustness and low power consumption [1]. A continuous demand for the higher memory capacity, better performance and more functionality at lower costs drives a continuously ongoing increase in the storage density, which is achieved by scaling of elementary memory devices. The current concept used in NVSMs is the floating-gate (FG) technology, in which the data is stored in form of electric charges within a conductive layer (floating-gate) embedded into a gate stack of a field effect transistor. Up to now the FG devices were able to scale like all silicon technology in accordance with Moore's law [2], which predicts a doubling of transistor density in an integrated circuit approximately every two-three years. The contemporary technological node for NAND architecture reached the 16 nm benchmark [3], resulting in several billions of transistors on one semiconductor chip. Further scaling will, however, be rather challenging, since the FG approach is already reaching its physical scaling limits [4], [5]. The inability to downsize the thickness of the insulating layers below 6 nm, the resulting high programming voltages, few-electron storage and cross-talk between neighbouring memory cells are the main obstacles for FG downsizing into 1x-nm and below [1], [6], [7]. In addition the FG cell exhibit several other drawbacks such as program/erase times in the range of microseconds to milliseconds, which makes it suitable only as a storage-type memory, as well as significantly higher operation voltages above 15 V in comparison to CMOS logic devices (below 1 V). Therefore, there is a strong demand on new memory concepts, which will be able to overcome the above mentioned limitations of the contemporary technology. The desired features are scaling potential below 16 nm, operation times of several nanoseconds comparable to volatile memories (SRAM and DRAM) in combination with non-volatile data storage possibility as well as low power operation. A wide variety of alternative memory approaches are being recently studied. Charge-trapping memories [8], [9], magnetic RAM [10], phase change RAM [11], resistive RAM [12] and ferroelectric RAM [13] are considered by the International Technology Roadmap for Semiconductors (ITRS) as the most promising candidates for future non-volatile memory applications and referred to as emerging memories [4].

In the ferroelectric memories the spontaneous polarisation of the ferroelectric materials is utilised for data storage. This type of memories provides the potential for fast operation due to polarisation switching capability in the nanosecond time range, fast unlimited endurance properties and non-volatile data storage. Moreover, since the ferroelectricity is an intrinsic material property with a single polarisation dipole stored within a unit crystal cell, the scaling can theoretically proceed down to the crystal unit-cell size [14]. The concept of a one-transistor ferroelectric cell (FeFET) is especially promising due to a cell design similar to the floating-gate memory but with a different gate stack structure. It provides the possibility for a non-destructive readout and a potential for a high integration density. The concept of a non-volatile ferroelectric transistor was proposed in the late 1950's [15]. Its industrial implementation is, however, still missing. The main obstacles originate from the conventional perovskite-type ferroelectric materials (lead zirconium titanate (PZT) and strontium bismuth tantalate (SBT)), in particular their integration [16], [17] and scaling issues [13]. The recently discovered ferroelectric properties of HfO_2-based thin films [18] – [20] renewed the interest in FeFET [21]. The decisive advantages of the HfO_2-based ferroelectrics are their full compatibility with the standard CMOS process and stable ferroelectric properties at film thicknesses in the nanometre range (5 – 30 nm) [22], [23]. The latter allow for a drastic reduction of the gate stack height, providing gate stack aspect ratios more suitable for scaling. The better scaling potential of HfO_2-based ferroelectrics is also assisted by a significantly lower dielectric constant of ~25 (for PZT or SBT ~200–300) and higher coercive field strength E_C of ~1 MV/cm (for PZT or SBT ~50 kV/cm). At reduced ferroelectric thickness these material properties enable to avoid high depolarization fields and compensate the memory window loss. Utilizing ferroelectric $Si:HfO_2$, FeFETs were fabricated at a state-of-the-art 28 nm technology node, which finally closed the scaling gap between the ferroelectric and CMOS logic transistors [24]. The effect of the ferroelectricity in HfO_2 has been recognised only several years ago. Therefore, there are still a lot of uncertainties about the origin of this ferroelectric behaviour as well as the impact of different fabrication conditions on its emergence. Moreover, the electrical behaviour of both, the HfO_2-based ferroelectric films and memory devices based on these films, requires more detailed studies.

The emphasis of this work lay on the properties of HfO_2 thin films, in which the ferroelectricity is induced by the silicon doping ($Si:HfO_2$). The potential and possible limitations of this material system with the respect to the application in non-volatile FeFET-type memories were extensively examined. Material aspects of $Si:HfO_2$ thin films were studied first using planar capacitor structures in order to get better insight into the ferroelectric properties and guidelines for transistor fabrication. The impact of the silicon doping concentration, post-metallisation annealing and film thickness on the emergence of ferroelectricity was investigated. By performing extensive electrical characterisation and structural analyses the correlation between the film crystalline structure and their electrical properties was found. The piezoelectric measurements were additionally carried out in order to confirm the structural ferroelectricity of the Si:HfO$_2$ films. Furthermore, their electrical properties that are relevant for memory applications, including the effect of the field cycling, polarisation switching kinetics and fatigue behaviour were investigated in detail. The performance of the Si:HfO$_2$-based FeFET memory devices, which were fabricated using the state-of-the-art 28 nm high-k metal gate CMOS technology, was investigated. The key memory characteristics, such as the program and erase behaviour, retention and endurance, were analysed. The impact of scaling of the Si:HfO$_2$-based FeFETs down to the gate length of 28 nm on their memory performance was investigated. The limited endurance and parasitic charge trapping were identified as the main issues of the studied devices. A detailed study of both these issues was performed in this work.

1 *Introduction*

2 Fundamentals

2.1 Non-volatile semiconductor memories

Non-volatile memories (NVMs) are a type of memory that enables information storage in the absence of an external power supply, in contrast to volatile memories, that lose the stored data when the power is turned off. NVMs include magnetic memories (e. g. magnetic tapes, hard drives), optical memories (optical discs) and non-volatile semiconductor memories (NVSMs) (e. g. NAND and NOR Flash, solid state drives). The main advantage of the NVSMs is that they are electrically addressable, which enables fast access times (1 ns – 50 μs). The absence of any mechanical parts makes them more robust and transportable than, for example, magnetic hard discs. Moreover, random access of each individual cell can be realised with NVSMs using the NOR memory architecture, a property that is not available for other memory types, in which the physical location of each individual data unit must be addressed mechanically. The price which NVSMs pay for this electrical control is a higher technological cost per bit and significantly lower storage densities in comparison to magnetic and optical memories. There are several types of NVSMs which are distinguished by their ability to be randomly accessed and rewritten [25]: mask-programmed read only memories (mask ROM), programmable read only memories (PROM), electrically programmable read only memories (EPROM), electrically erasable and programmable read only memories (EEPROM) and Flash EEPROM. For example, EEPROM type cells can be electrically erased and programmed byte-wise. The Flash EEPROM type cells can be also electrically erased and programmed; the erase is, however, performed on large blocks, whereas programming remains a byte wise operation. The Flash EEPROM, which enables the highest data storage density, is currently the leader in the NVSM market. The basic structure of all modern electrically programmable NVSMs is a field effect transistor with a floating gate.

2.1.1 Metal-insulator-semiconductor field-effect transistor

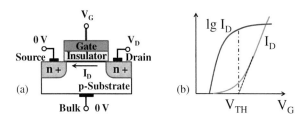

(a) (b)

Figure 2.1 (a) Schematic diagram of an n-channel MISFET and (b) its transfer characteristic with drain current (I_D) or logarithm of I_D plotted versus gate voltage (V_G).

The metal-insulator-semiconductor field effect transistor (MISFET) forms the basic structure of non-volatile semiconductor memories. Figure 2.1 (a) shows a schematic drawing of an n-channel MISFET device. It consists of a p-type semiconductor substrate with two highly doped n-regions, the source and drain. The conductivity of the channel between source and drain regions can be capacitively controlled by applying a voltage to the gate electrode, which is separated from the semiconductor surface by an insulator. If a negative voltage is applied to the gate electrode, the majority charge carriers, holes, are drawn from the bulk of the semiconductor substrate and accumulate on the surface, resulting in a low electron conductivity of the channel. A positive gate voltage, on the other hand, repels the holes and attracts the electrons. If the applied positive gate voltage is high enough, it leads to the inversion of the conductance type at the semiconductor surface. An n-conductive channel is formed, which enables electrons to flow between the source and drain regions. The gate voltage required to build up an electron conductive channel is called the threshold voltage (V_{TH}). For an n-channel MISFET the threshold voltage can be calculated as [25]:

$$V_{TH} = V_{FB} + 2\psi_B + \frac{\sqrt{2\varepsilon_S \varepsilon_0 q N_A \cdot (2\psi_B)}}{C_i} =$$
$$= \left(\Phi_{MS} - \frac{Q_{eff}}{C_i} \right) + 2\psi_B + \frac{\sqrt{2\varepsilon_S \varepsilon_0 q N_A \cdot (2\psi_B)}}{C_i},$$

(2.1)

where q is the unit electronic charge, ε_0 – the permittivity in vacuum, ε_S – the relative semiconductor permittivity (for Si $\varepsilon_S = 11.9$), N_A – the acceptor concentration of the p-type semiconductor substrate, ψ_B – the Fermi potential, which is the Fermi level energy with respect to the middle of the semiconductor band gap, and C_i – the capacitance of the gate insulator or gate insulator stack in case of multilayer structures. The threshold voltage comprises three terms: (1) the flatband voltage (V_{FB}) required to achieve flatband conditions at the semiconductor surface with zero surface potential $\psi_S = 0$, (2) a voltage of $2\psi_B$ required to set the semiconductor surface into a strong inversion (where $\psi_S = 2\psi_B$) and (3) the voltage required to compensate the charge of the depletion region of the semiconductor substrate (last term in (2.1)). The value of the flatband voltage is determined by the work function difference between the metal of the gate electrode and the semiconductor (Φ_{MS}) as well as by the voltage drop, caused by the charges within the insulating layer. The insulator charges are characterised by an effective net surface charge density (Q_{eff}), calculated under the assumption that all charges are located at the semiconductor-insulator interface. From the definition of the threshold voltage in (2.1) it can be seen that V_{TH} value is affected by the properties of the semiconductor substrate, e.g. its doping concentration, as well as properties of the gate stack such as the insulator capacitance and the electrode work function. Special attention should be drawn to the fact that an introduction of charges into the insulator layer of the gate stack provides an additional possibility to adjust the threshold voltage. This property is decisive for the NVS memories, as will be discussed in the next chapter 2.1.2.

If a voltage is applied to the drain electrode (V_D) while setting the source and bulk contacts to the ground potential, a drain current (I_D) flows through the transistor channel. The magnitude of I_D can be controlled by the applied gate voltage (V_G). The transistor can be operated in different modes (linear, non-linear and saturation) depending on the relationship between the gate and drain voltages [25]. In the linear mode, with $V_D \ll (V_G\text{-}V_{TH})$, where the conductive channel can be treated as a uniform resistive layer, the drain current yields [25]:

$$I_D = \frac{W}{L}\mu_n C_i \left(V_G - V_{TH} - \frac{V_D}{2} \right) \cdot V_D, \tag{2.2}$$

where W and L denote the width and length of the transistor channel, respectively, and μ_n stands for electron mobility in the channel. The dependence of the drain current on the gate voltage at constant drain bias is referred as the transistor transfer characteristic ($I_D\text{-}V_G$) (Figure 2.1 (b)). Above the threshold voltage a conductive channel is formed and the drain current exhibits a linear dependence on the gate voltage. The V_{TH} value can be experimentally obtained by extrapolating the linear region of the $I_D\text{-}V_G$ curve to the zero drain current value or by using a constant drain current criterion [26]. The region below V_{TH} is called the subthreshold region. Here the semiconductor surface is in weak inversion or depletion and the drain current is dominated by electron diffusion along the channel due to the electron-density gradient and not by the drift in the lateral electric field. In the subthreshold region the drain current exhibits an exponential dependence on the gate voltage.

2.1.2 Floating-gate memory technology

In the present age of digital electronics, information is stored in form of digital states. The main principle of a data storage device (memory cell) is the possibility to realise of at least two stable states, which represent logical "0" and "1". In this way 1-bit digital information can be stored by a single memory cell. These two states should be retained independent of an external power supply to enable non-volatile operation.

The modern NVSMs are charge-based devices, where the data is stored in the form of an electric charge within a field effect transistor with a floating-gate. The structure of a floating-gate (FG) cell is schematically illustrated in Figure 2.2 (a). It consists of a field effect transistor as a basic structure with an additional conductive layer, commonly lightly doped polysilicon, embedded into the insulator layer of the gate stack. This conductive layer is called the floating-gate, since it is completely surrounded by an insulator and its potential is floating. The second electrode, control gate (CG), is used for the external control of the channel conductivity. The memory operation is realised by storing the charges on the floating-gate, which is electrically isolated and, thus, retains its charge even if the power is switched off. The conductance of the transistor channel is directly affected by the presence of charges on the floating-gate. For example, the stored electrons will lower the channel conductivity, resulting in a decreased drain current for the same voltage on the control gate (as shown in Figure 2.2 (a) by the thickness of the I_D arrow). The V_{TH}-value of the transistor increases according to (2.1), leading to a positive shift of the I_D-V_G characteristic. Thus, at least two memory states can be distinguished for FG cell: "ON"-state, corresponding to high channel conductivity and low V_{TH}-value, and "OFF"-state, corresponding to low channel conductivity and high V_{TH}-value ((Figure 2.2 (b)). The read operation can be performed non-destructively by sensing the drain current.

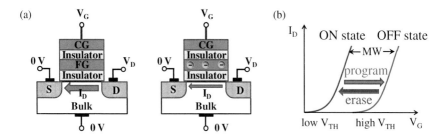

Figure 2.2 Floating gate cell and its data storage principle. (a) Two memory states of a FG cell with and without electrons stored on the FG electrode, representing the "ON"- and "OFF" states, respectively. (b) I_D-V_G characteristics in two memory states.

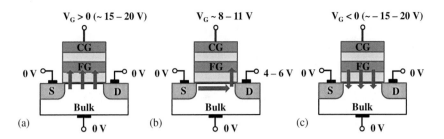

Figure 2.3 Program and erase operations of a FG cell. (a) Program by Fowler-Nordheim tunneling, (b) program by channel hot-electron injection, (c) erase by Fowler-Nordheim tunneling.

The state of the FG cells can be rewritten by changing the amount of charge stored on the floating-gate. The injection of electrons onto the floating-gate is commonly referred to as the program operation which sets the cell into the "OFF" state. During the inverse erase operation the electrons are removed from the floating-gate and the cell is returned into the initial "ON" state. Both program and erase can be performed electrically due to the presence of the control-gate electrode. Two mechanisms for electron injection can be exploited to program the FG cell [27], [1]: Fowler-Nordheim (FN) tunneling and channel hot-electron injection (CHEI). The voltages applied to the transistor terminals and the accompanying electron flow are shown in Figure 2.3 for each approach. For programming with Fowler-Nordheim tunneling (Figure 2.3 (a)) a positive voltage is applied to the control gate and all other terminals are kept at 0 V. Electric fields of 8 – 10 MV/cm are required across the bottom insulator to induce charge injection from the substrate into the floating-gate. Both insulating layers of the gate stack should be at least 6 nm thick in order to ensure good retention of the electrons stored on the floating-gate. Therefore, high operation voltages of about 15 – 20 V have to be used to achieve the required injection fields. In the case of Fowler-Nordheim tunneling the electrons are injected onto the floating-gate uniformly along the channel. The efficiency of this program operation depends on the gate coupling ratio (GCR), which is defined as the ratio of the CG-to-FG capacitance to the total FG capacitance and determines the voltage distribution within the gate stack [27]. The GCR must be more than 0.6 in order to ensure a predominant voltage drop across the bottom insulating layer. The mechanism of FN tunneling is rather slow, resulting in operation times in the range of milliseconds. On the other hand, using channel hot-electron injection (Figure 2.3 (b)), the programming can be accomplished in several microseconds. In this case relatively high drain voltages are applied (4 – 6 V) to accelerate the electrons in the lateral electric field to the energies sufficient to overcome the potential barriers between the channel and the floating-gate. The injection takes place predominantly near the drain region, where the electrons exhibit the maximum energy. Two types of electrons contribute to the injection current: electrons accelerated in the channel, scattered by the lattice and redirected towards gate as well as electrons generated by impact ionisation that gained the required energy. In order to make the programming more efficient, a positive voltage (8 – 11 V) is applied to the control-gate. This approach is rather power consuming, since high currents flow in the channel. The erase operation is performed via Fowler-Nordheim tunneling by applying a negative voltage (about – 15 – 20 V) to the control-gate or positive voltage to the source/drain, which induces back-tunneling of the stored electrons from the floating-gate into the channel (Figure 2.3 (c)).

Tremendous success in the development of the FG memory technology has been achieved in the last several decades. The compact design of the memory cell consisting of a single transistor enabled its continuous rapid scaling and, thus, facilitated a continuous increase of the data storage density. The NVSM market is dominated currently by the NAND-type flash memories with which the highest storage density with commercially available products up to 128 Gb is achieved. Further increases of the storage density will, however, be rather challenging, since the FG approach is already reaching its scaling limits. The contemporary technology node is 16 nm [3], whereas already below 15 nm node new memory concepts will be required according to the International Technology Roadmap for Semiconductors [4]. The main obstacles for FG scaling are the inability to downsize the thickness of the insulating layers below 6 nm and the resulting high programming voltages, the need to maintain a high gate coupling ratio, retention issues due to few-electron storage and cross-talk between neighbouring memory cells [1], [6], [7]. Other drawbacks of the FG cell concept are the relative slow programming times in the range of microseconds to milliseconds as well as endurance restricted to a maximum of $10^4 - 10^6$ program/erase cycles due to the intrinsic charging and degradation of the bottom insulating layer exposed to repeated charge injection.

2.2   Emerging memory concepts

Several new alternative concepts, referred to as emerging memories, were proposed as possible replacement for the conventional FG approach. The most promising candidates for future memory application are charge-trapping memories, magnetic memories, phase change memories, resistive memories and ferroelectric memories. A short overview of the first four concepts, including their operation principle, advantages and drawbacks, will be given in this chapter. The ferroelectric based memories will be discussed in detail in chapter 2.3.

Charge-trapping (CT) memories that store the charge on localised trap states within a dielectric layer instead of a conductive floating-gate are regarded as the most probable successor to the FG approach for the technology nodes below 15 nm [4]. Silicon nitride commonly serves as the charge storage layer. The charge-trapping memories are also called

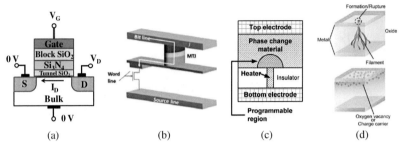

Figure 2.4 Emerging memory concepts: (a) SONOS-type charge-trapping memory, (b) magnetic RAM [28], (c) phase change RAM [11] and (d) resistive RAM [29].

SONOS-type memories because of the layer sequence used to form the transistor gate stack: polysilicon gate, block oxide, silicon nitride and tunnel oxide on a silicon substrate (Figure 2.4 (a)). The charge is stored in the localised trap sites so this memory type is more tolerant to tunnel oxide imperfections in comparison to the FG approach. Therefore, thinner oxide films (3 – 4 nm) can be utilised, resulting in a reduction of the operation voltages. Moreover, charge-trapping cells do not suffer from the gate coupling ratio issue, since the channel is directly controlled by the control gate, and the cross talk between the neighbouring cells is significantly reduced. The main weak point of SONOS cells is their limited erase capability, which can be improved only at the expense of the retention properties. Proposed solutions to this problem include band engineering of the tunnel oxide (BE-engineering) [30] and implementation of high-*k* dielectrics for the block oxide in combination with high workfunction metals (MANOS cells) [31]. Moreover, since the change-trapping memories exploit the program/erase concepts similar to the FG, they experience the same scaling obstacles: word-line to word-line breakdown due to the high operation voltages and retention degradation as a result of few electron storage. Therefore, planar scaling below 15 nm will also be challenging for the charge-trapping memories. On the other hand, alternative solutions for increasing the storage density become feasible, such as multi-bit storage in the NROM cells [8], [32] and 3D architectures [9], [33], which enable higher storage densities without the need to decrease the cell size. Ultra-high storage density can be achieved using the latter approach. Complication of the fabrication process is, however, a consequence.

Magnetic random access memories (MRAMs) exploit the tunnel magneto-resistive effect [34] for data storage and belong to the class of non-charge-based non-volatile memories. The elementary cell consists of a magnetic tunnel junction, two ferromagnetic thin films separated by an insulating tunnel barrier, and an access transistor connected in series (Figure 2.4 (b)). The memory states differ in the resistivity of the junction, which is determined by the relative orientation of the magnetisation in the ferromagnetic layers. A low resistivity state corresponds to the parallel magnetisation orientation, whereas antiparallel orientation induces a high resistivity state. The magnetic junctions always include one pinned ferromagnetic layer with fixed orientation of magnetisation and the second free ferromagnetic layer with a changeable orientation of magnetisation. Utilisation of the spin-transfer torque [10], [35] for magnetisation switching enables realisation of highly scaled structures and significantly lowers the power consumption during write operation. Magnetic RAM has already been demonstrated many universal memory features: fast read and write in range of several nanoseconds comparable to DRAM or even better and just slightly worse than SRAM, low operation voltages of around 2 V and an almost unlimited endurance of 10^{12} write/read cycles [4]. The main obstacle is scaling below 30 nm. The reduction in the cell size degrades the thermal stability, resulting in poor retention. The solution can be found in introduction of new materials [28], which are still at the development stage. Moreover, the complicated structure of the magnetic tunnel junction, including 10 – 12 different layers with thickness of 0.8 – 2 nm, makes the fabrication process challenging from the viewpoint of deposition and etching.

Phase change random access memories (PCRAMs) are resistive-type memories similar to the MRAMs, where the memory states differ in their resistance. The basic memory cell is also similar to the MRAM cell (Figure 2.4 (b)). It consists of a resistor with an adjustable resistance and an access device, typically a field effect or bipolar transistor. The variation in resistance in PCRAM is based on the reversible transition between a high resistive amorphous phase and the low resistive crystalline phase of the functional material. The chalcogenide alloys (most commonly germanium-antimony-tellurium Ge-Sb-Te alloys) are used as the storage medium. These materials have already found wide application in the optical-type memories (CD and DVD disks), which make use of the difference in optical properties for the amorphous and crystalline phases. The PCRAMs utilise, on the contrary, the dependence of the electrical resistance of these materials on their crystalline structure. The resistor of a PCRAM cell consists of top and bottom electrodes with the phase change layer and the heater resistor material embedded in between (Figure 2.4 (c)). The phase change occurs in the local volume of the chalcogenide film above the heater and is accomplished by applying a current pulse, which causes local heating of a material. The amplitude and the width of the current pulse determine the resulting memory state (crystalline structure). A short pulse of high amplitude establishes the high-resistive amorphous state, whereas a longer pulse of lower amplitude is required to obtain the low-resistive crystalline state [11]. The main advantages of this memory type are fast switching times (10 -100 ns), low operation voltages (1 – 2 V), high endurance with 10^9 cycles [4], the capability of multi-level data storage [11] and relatively uncomplicated integration in to the CMOS process flow. The main obstacles arise from the requirement of high current density during the writing of a memory state. It limits the scaling of the access device and, thus, the entire memory cell [36]. Moreover, high current densities can cause a deterioration of the endurance characteristics [36]. Other matters of concern are the thermal cross-talk between the cells at high storage densities, the read disturb and the retention degradation caused by the structural relaxation [1].

Resistive random access memories (ReRAMs) are another memory type that exploits a reversible change in the resistance of a metal-insulator-metal structure as the data storage principle. The resistive switching is based on electrically induced ionic motion combined with reduction/oxidation (redox) electrochemical reactions. Therefore, this memory type is also often referred as Redox RAM. The memory cell consists of a resistor, which is typically a metal-insulator-metal capacitor that exhibits resistive switching behaviour, and an access device such as a field effect transistor or a diode. There is a wide variety of materials that demonstrate resistive switching [12]: binary transition metal oxides (TiO_2, Nb_2O_5), perovskite-type complex transition metal oxides materials ((Ba, Sr)TiO_3, Pb(Zr_xTi_{1-x})O_3), large band gap high-k dielectrics (Al_2O_3), chalcogenides and organic compounds. Further, the ReRAM cells differ in their operation principle. A classification [12], [37] is possible in terms of the electrical switching behaviour (unipolar and bipolar), the conductance mechanism (filament-type and interface-type conductance (Figure 2.4 (d))) or the physical mechanisms responsible for switching (thermo- chemical mechanism, valence change mechanism and electrochemical metallisation). The ReRAMs are able to provide fast read and write operation speeds in the nanosecond time range (<10 ns) at low operation voltages (< 1 V). They are considered to be promising candidates for high-density integration due to the potential scalability of the cell below 10 nm [38], multi-bit storage and the functionality of crossbar-array architectures that use passive access devices (diodes) instead of transistors. Although high endurance (> 10^{10} write cycles) and good retention properties have been demonstrated for some material systems [39], [40], little data on the statistical analyses is available. Reliability issues such as endurance and retention cannot be excluded, since none of the cell resistance states can be considered as thermodynamically stable [12]. An understanding of the driving force for the resistive switching remains unclear, which is the main obstacle for ReRAM commercialisation that requires a well-understood, predictable and stable technology.

2.3 Ferroelectric memories

2.3.1 The ferroelectric effect

Ferroelectricity is a particular property of some dielectric materials in which they exhibit a spontaneous electric polarisation that can be reversed by an external electrical field. The materials that reveal these properties are called ferroelectrics. Typical examples of widely used ferroelectric materials are barium titanate (BTO) [41] lead zirconate titanate (PZT) [42], [43], [44] and strontium bismuth tantalate (SBT) [14], [45]. Ferroelectric properties have also been found in several organic materials [46], [47].

The ability of a crystal to show ferroelectric behaviour is directly determined by its crystallographic symmetry. Only non-centrosymmetric crystals, which have no inversion centre, can possess ferroelectric properties. In this case the unit cell is allowed to have unique crystallographic directions not mirrored by any symmetry element. This crystallographic direction can act as a polarisation axis. Figure 2.5 (a) illustrates two stable configurations of

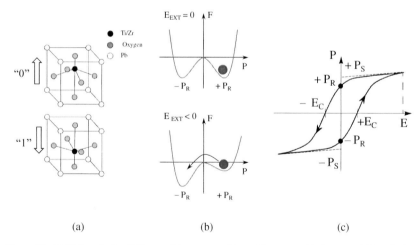

(a)   (b)   (c)

Figure 2.5 (a) Unit cell of a PZT crystal in two possible stable states with the central Ti or Zr ion displaced toward either the upper or lower oxygen ion along the tetragonal symmetry axis giving rise to a net electric polarisation directed up (state "0") or down (state "1") [49]. (b) Free energy as a function of polarisation for ferroelectric materials without or with an applied external electric field E_{EXT}. (c) Typical ferroelectric hysteresis loop showing polarisation (P) as a function of external electric field (E). The three important loop parameters are: spontaneous polarisation (P_S), remanent polarisation (P_R) and coercive electric field (E_C).

the unit cell of PZT crystal in the tetragonal ferroelectric phase.The central ion (Ti^{4+} or Zr^{4+}) is displaced from the central position toward either the upper or lower oxygen ion O^{2-}so that the centres of the net negative and positive charges are shifted relative to each other. The net electric polarisation is directed either up or down, which can represent the two logical states "0" and "1". These two thermodynamically stable atom configurations can be described by means of a double-well potential with two equilibrium positions corresponding to the same minimum value of the free energy and separated by a potential barrier [48] (Figure 2.5(b)). At any given time the central ion of a unit cell (Ti^{4+} or Zr^{4+}) is located at one of the two possible positions corresponding to one of the energy minima. In the presence of an external electric field the potential barrier is lowered so that the ion can change its position and jump into the other potential minimum. Thus, the polarisation direction is reversed. In ferroelectric materials the relationship between polarisation and electric field is represented by the hysteresis loop (Figure 2.5(c)). It can be used to extract the characteristic material parameters: the spontaneous polarisation (P_S), the remanent polarisation (P_R) and the coercive electric field (E_C). P_S is obtained from extrapolation of the saturating linear part of the hysteresis to zero field value, whereas P_R is the actual polarisation remaining in the crystal after the electric field has been removed. E_C corresponds to the value of the external electric field with a polarity opposite to the remanent polarisation that is required to reduce the latter to zero. P_R and E_C are the decisive characteristics of ferroelectric materials for memory application and determine the operation voltage and memory window. Table 2.1 summaries these characteristics for the most important ferroelectric thin films.

The local crystal regions with the same orientation of spontaneous polarisation are referred to as ferroelectric domains. The transition regions between domains with different polarisation directions are called domain walls and are $1 - 10$ nm thick comprising only $2 - 3$ unit cells [54], [55]. In ferroelectric materials a multi-domain structure is commonly formed in order to minimise the total crystal energy [56], [49]. This process is determined by the electrostatic energy associated with the depolarisation field that arises from non-compensated

Table 2.1 Key characteristics of typical ferroelectric thin films [49] – [52].

Materials	P_R ($\mu C/cm^2$)	E_C (kV/cm)	ε
Pb(Zr,Ti)O$_3$ (PZT)	25-35	50-70	300-1300
SrBi$_2$Ta$_2$O$_9$ (SBT)	10-25	30-50	120-250
BaTiO$_3$ (BTO)	3-15	30-50	300-1000
(Bi,La)$_4$Ti$_3$O$_{12}$ (BLT)	15-20	80-100	150-300
Polymer ferroelectrics	2-10	50-5000	10

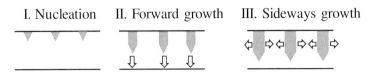

I. Nucleation   II. Forward growth   III. Sideways growth

Figure 2.6 Mechanisms of polarisation switching in ferroelectric films after [53].

polarisation charges, as well as the elastic energy associated with the mechanical constraints. The impact of mechanical stress on the formation of the domain structure is especially important in polycrystalline materials and thin films [56], [49].

Domains with different polarisations are statistically distributed in unpoled ferroelectric materials resulting in a zero net polarisation. In the presence of an external electric field, the domains reverse so that the polarisation direction of most domains coincides with that of the external field. Three regimes for the switching of the polarisation in ferroelectric films are generally distinguished [53]: nucleation of the domains with polarisations similar to the external field, forward growth of needle-like domains in the direction of the external field and sideways domain expansion (Figure 2.6). The two latter processes occur by means of domain wall motion. Nucleation commonly takes place at the electrode interface or at grain boundaries, where the formation energy is the lowest. The time needed for a nucleus to reach the critical size required for further growth is called the nucleation time, which is about 1 ns for oxide ferroelectrics [53]. The time for forward growth ($t_{FG} = d/v$) is determined by the film thickness (d) and the speed of sound (v). For films thinner than 1 µm t_{FG} lies in the picosecond time range and can be neglected. The time for sideways domain growth depends on the strength of the applied electric field (E) according to $t_{SG} = C \cdot E^{-3/2}$, where C is a constant. The rate-limiting mechanism depends on the material properties, film thickness [57], crystalline structure (monocrystalline or polycrystalline) [16], temperature [58] and lateral cell size [53]. There are two main models that describe the polarisation switching process [16]: (1) the Kolmogorov-Avrami-Ishibashi model developed by Ishibashi and Tagaki [59], [60], which treats the polarisation reversal in terms of domain wall motion, and (2) the nucleation-limited switching model proposed by Tagantsev [61], which considers the nucleation of reversed domains as the limiting switching mechanism. The first model is applicable for single crystals [62] and epitaxial films [63], whereas the second model describes the switching behaviour in polycrystalline films [61], [64]. Switching typically proceeds in the nanosecond time range (2 – 200 ns) for ferroelectric thin films [65] – [67].

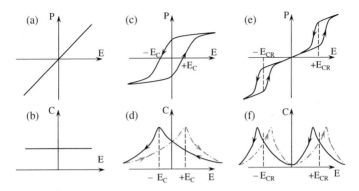

Figure 2.7 Characteristic form of the polarisation-electric field *P(E)* and capacitance-electric field *C(E)* curves for three material classes: (a), (b) ideal linear dielectrics, (c), (d) ferroelectrics and (e), (f) anti-ferroelectrics.

A hysteretic dependence between polarisation and electric field is a necessary but not a sufficient condition of true ferroelectricity in a material [68], [69]. Several other artefacts, such as surface polarisation, trapping-detrapping phenomena at Schottky-like electrodes [70] or leakage currents [71], can be also responsible for an experimentally detected hysteresis loop. Additional information about the true material properties can be gained by analysing the voltage dependence of the small signal capacitance. The polarisation-field and small signal capacitance-field curves characteristic for three classes of materials: linear dielectrics, ferroelectrics and antiferroelectrics, are shown in Figure 2.7. In case of linear dielectrics the electric polarisation exhibits a linear dependence from the electric field and field-independent capacitance value (Figure 2.7 (a), (b)). Polarisation hysteresis loops and butterfly-shaped capacitance-field curves are characteristic for ferroelectric materials [72], [73] (Figure 2.7 (c), (d)). The response measured during the capacitance-voltage test is associated with ionic and electronic displacements as well as with reversible domain wall motions around the local energy minima [72]. Since the domain wall concentration is highest at the coercive fields, the value of capacitance should peak at these points [72]. Antiferroelectric materials are similar to ferroelectrics in that they possess spontaneous dipoles originating from their crystal structure. The main difference is, however, that the neighbouring dipoles are aligned antiparallel to each other, resulting in a zero net polarisation. Antiferroelectrics can be field-forced to undergo a phase transition into the ferroelectric phase by applying an external electric field, when a critical field value (E_{CR}) is achieved [74], [75]. This ferroelectric phase is, however, unstable and transforms back into the antiferroelectric phase when the external field is removed. Therefore, antiferroelectric materials do not exhibit any remanent polarisation. This possibility of field-induced phase transition in antiferroelectric materials leads to a double-loop polarisation hysteresis and double-butterfly-shaped capacitance-field curve (Figure 2.7 (e), (f)).

True ferroelectric materials should also reveal piezoelectric and pyroelectric properties in addition to a polarisation hysteresis. This follows from the classification of dielectric materials (Figure 2.8). Ferroelectrics are a special subclass of dielectric materials that belong to pyroelectrics. These are, in turn, a special subclass of piezoelectric materials. The effect of pyroelectricity is based on the temperature dependence of the spontaneous polarisation of the crystal. In a pyroelectric crystal the change in the net polarisation is proportional to the temperature change, which can be measured in a closed circuit as a current flow or in an open circuit as a voltage change across the crystal. Piezoelectricity is the property of a material to acquire an electric polarisation in response to the applied mechanical stress. This is referred as direct piezoelectric effect. The converse piezoelectric effect consists in inducing the deformation of a sample by applying an external electric field. Expansion/constriction is detected when the external electric field is parallel/antiparallel to the direction of the existing spontaneous polarisation in the sample. In ferroelectric materials the displacement versus electric field dependence has the so-called butterfly shape [77], [56], which is caused by polarisation reversal. Figure 2.9 (a) shows the correlation between the idealised polarisation and displacement loops of a ferroelectric, in which the polarisation reverses by 180° [56], [76]. In real ferroelectric materials the shape of the displacement loop is smoother (Figure 2.9 (b)) due to the distribution in the domain coercive fields and existence of non-180° domains [56]. The converse piezoelectric effect is utilised in piezoresponse force microscopy (see chapter 3.2.4) for the visualization of ferroelectric domains.

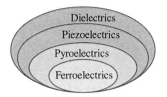

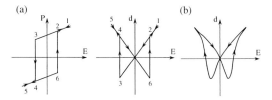

Figure 2.8 Classification of dielectric materials. Ferroelectrics are a specific subclass of dielectric materials and must possess properties of all upper classes.

Figure 2.9 Converse piezoelectric effect [76]. (a) Correlation between the idealised polarisation *P(E)* and displacement *d(E)* loops in a ferroelectric material, in which the polarisation reverses by 180°. (b) *d(E)* loop shape of real ferroelectric materials.

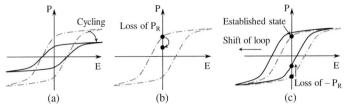

Figure 2.10 Schematic illustration of the main ferroelectric degradation mechanisms after [49]: (a) fatigue, (b) retention loss and (c) imprint.

The reliability properties of ferroelectric materials are essential for their application in memory devices. Three main degradation mechanisms are commonly discussed: fatigue, retention loss and imprint. Each of them is accompanied by a specific change in the polarisation loop (Figure 2.10). The fatigue effect is a result of repetitive polarisation reversal, which causes flattening of the polarisation loop and lowering of the switchable polarisation (Figure 2.10 (a)). Several theoretical models explaining fatigue phenomenon have been developed for perovskite-type ferroelectrics [53], [78] – [82]. The degradation of the switchable polarisation is attributed to modification of the switching process during cycling, where either the domain walls get pinned by mobile charged defects [82] – [84] or the growth of domain nuclei with opposite polarity becomes inhibited [78] – [80]. Two main microscopic origins of fatigue have been proposed – oxygen vacancies [79], [85], [86], redistributing within the ferroelectric layer under electrical stress, or free charges injected from the electrodes [78], [82], [87] – [89]. Retention loss is characterised by a decrease of the polarisation with time in cells with a primary established polarisation state (Figure 2.10 (b)). The retention properties are decisive for non-volatile type memories, where the capability of information storage for at least 10 years is required [1]. The depolarisation field, internal built-in bias and charge injection are considered to be the main driving forces for polarisation loss [90] – [92]. The depolarisation field arises from incomplete compensation of the ferroelectric polarisation at the electrode interfaces. This is the case for electrodes with low free charge carrier density [91], where a depletion layer appears, or for an insulating layer with low dielectric constant, which is embedded between the electrode and the ferroelectric layer [93]. The depolarisation field is an essential issue for metal-ferroelectric-insulator-semiconductor (MFIS) stacks that can lead to retention loss. If one of the polarisation states is retained for a long time, a build-up of the internal bias can take place due to the redistribution of mobile charges within the ferroelectric layer or charge injection through the interfacial layer [16], [81]. This effect is called imprint. It is characterised by a shift of the polarisation hysteresis loop along the E-axis (Figure 2.10 (c)). The imprint effect leads to polarity dependent retention behaviour. The internal bias stabilises one of the polarisation states while impairing the stability of the opposite state.

2.3.2 Types of ferroelectric memories

Ferroelectric materials can be used as an information storage media due to their ability to switch between two stable polarisation states. Progress in the fabrication of thin ferroelectric films has led to the development of two basic types of ferroelectric memories: **F**erroelectric **R**andom **A**ccess **M**emory (FeRAM) [42] and **F**erroelectric **F**ield **E**ffect **T**ransistor (FeFET) [14], [94]. These two concepts differ in the structure of the elementary memory cell and the readout approach. In the FeRAM the ferroelectric layer is integrated into the capacitor of a DRAM-like cell, resulting in a one transistor-one capacitor (1T-1C) memory cell. The FeFET-type cell consists of a single transistor (1T) with a ferroelectric layer built directly into the transistor gate-stack. The main advantages of ferroelectric memories are non-volatility, fast read/write times (under 50 ns), low operation voltages, very high endurance (greater than 10^{15} write/read cycles) and low power operation [16], [1], making ferroelectric memory especially attractive for mobile applications.

The FeRAM concept is a more developed one with commercial products already available including embedded (e.g., RFID and microcontrollers) and stand-alone applications (e.g., smart cards). The ferroelectric materials most widely utilized in modern FeRAM cells are PZT [42], [95] and SBT [45], [96] films. One of the main drawbacks of this memory type is a destructive readout scheme, where the polarisation of the cell is switched in order to sense the stored memory state. A voltage pulse is applied to the capacitor during reading and a transient current response is simultaneously sensed. Depending on the initial polarisation state the ferroelectric polarisation either is reversed or remains unchanged, resulting in different value of the transient current response. Since the cell state is changed during the readout operation, it must be rewritten each time after reading. This imposes the requirement of a high endurance resistivity on the ferroelectric material used. Another important issue is the incompatibility of conventional ferroelectric compounds with standard CMOS technology. The main FeRAM integration challenges include [97], [98], [17]:

- interdiffusion of ferroelectric oxides' constituents and silicon, resulting in a performance degradation of both the ferroelectric capacitor and the transistor,
- high processing temperatures of ferroelectric films, promoting interdiffusion and impacting the doping profile of the transistors,
- oxidation of the interconnect metal layers due to the high pressure oxygen atmosphere used during the fabrication of the ferroelectric films,
- incorporation of hydrogen into the ferroelectric films during the forming gas annealing step, impairing their ferroelectric properties,
- etching during structuring of ferroelectric capacitors.

In addition FeRAM, with its large cell size, can hardly compete with the conventional FG technology in the field of storage density and cost per bit. Since a FeRAM cell includes at least one transistor and one capacitor, it is difficult to scale it along with the CMOS technology roadmap. FeRAM has already faced its scaling limit at a contemporary node of 130 nm [99]. The challenge of further cell miniaturisation originates from the minimum signal level (total capacitor charge) required for sensing [100], which is directly proportional to the capacitor area.

The concept of the FeFET-type memory, where the insulating layer in the gate stack of a standard field effect transistor is replaced by a ferroelectric film, was proposed in the late 1950's [15]. Figure 2.11 schematically illustrates the structure of a FeFET memory cell and its basic operation principle using the example of an n-channel device. The conductivity of the transistor channel is modulated by the polarisation charge of the ferroelectric layer, which can be controlled by a voltage applied to the gate electrode. It should be noted that the effect of the gate voltage on the channel conductivity in FeFETs is opposite to that of charge-trapping memories (chapter 2.1.2). A positive gate voltage results in a positive polarisation charge at the ferroelectric-semiconductor interface, attracting electrons and increasing the channel conductivity. In this case the FeFET transistor is in the "ON" state with the I_D-V_G characteristic shifted to lower gate voltage. An "OFF" state is induced by applying a negative gate voltage which is sufficient to reverse the ferroelectric polarisation. The conductivity of the channel decreases due to the negative polarisation charge at the ferroelectric-semiconductor interface and the I_D-V_G curve shifts to higher gate voltage. The readout of this cell can be performed non-destructively by sensing the drain current without changing the polarisation of the ferroelectric layer. This relaxes the requirement for unlimited endurance stability and lowers the power consumption. Furthermore, the compactness of a one transistor (1T) cell provides better scaling potential. Since the surface charge density and not the total

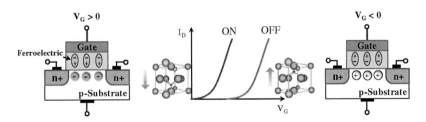

Figure 2.11 Basic structure and operation principle of a FeFET-type memory cell.

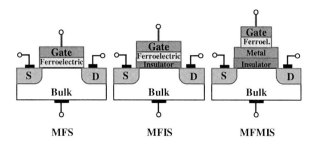

Figure 2.12 Three possible implementation designs of a FeFET-type memory cell.

charge, as for FeRAM, is decisive during sensing the conductivity of the transistor channel, the FeFET-type cell has a potential to be scaled along with the CMOS technology.

Despite the simplicity of the 1T cell idea, its physical realisation has turned out to be challenging, which has prevented its industrial implementation up until now. The main obstacles originate from the integration [16], [17] and scaling issues [13] of conventional perovskite-type ferroelectric materials such as PZT and SBT. In a FeFET-type cell the ferroelectric material is in direct proximity to the silicon interface, in contrast to the FeRAM cell, in which the capacitor and transistor are two physically independent devices. Therefore, for the 1T ferroelectric cell the incompatibility issues of the typical ferroelectric materials with conventional CMOS transistor technology becomes even more crucial. Interdiffusion of silicon and components of the ferroelectric oxides' impairs the ferroelectric properties [97], [17]. Simultaneously, the silicon interface degrades, resulting in increased interface trap density. Moreover, due to the direct contact between ferroelectric material and semiconductor channel, charge injection from the channel becomes inevitable, compensating the polarisation charge and leading to poor retention properties. In fact, the charging trapping was so severe in the first FeFET devices [101], [102] that it dominated the V_{TH} shift after write operation and masked the ferroelectric polarisation. The introduction of a high-k interface buffer layer between the ferroelectric and semiconductor was required to solve the interdiffusion and charge-trapping issues [103] – [105], resulting in a metal-ferroelectric-insulator-semiconductor (MFIS) cell type (Figure 2.12 centre). Further, HfO_2-based buffer layers have enabled the poor retention to be overcome [105], [106], which was a serious matter of concern for a long time. A significant drawback of the perovskite-based ferroelectric transistors is their scaling limitation is another, which is caused by the necessity to grow thick ferroelectric films (200 – 500 nm) [13]. This is required to ensure an adequate voltage drop across the ferroelectric layer, to enable long data retention times and to provide a sufficient memory window (*MW*) between the two memory states. The *MW* can be estimated, to a good approximation, as twice the product of the coercive field (E_C) and thickness of the ferroelectric layer (d_{FE}) [107], [108]:

$$MW = V_{TH}^{OFF} - V_{TH}^{ON} \cong 2E_C \cdot d_{FE}. \tag{2.3}$$

The requirement for a sufficiently large MW imposes a limitation on the scaling of the ferroelectric film thickness. The lowest tolerable d_{FE} depends on the E_C-value, which is a characteristic material parameter. In FeFETs utilising conventional ferroelectrics (PZT and SBT) with relative low coercive fields of about 50 kV/cm (Table 2.1), at least 100 nm thick films are needed to obtain a memory window of 1 V. The reduced voltage drop across the ferroelectric layer and increased depolarisation field, which impair the retention properties, are both shortcomings of the MFIS-FET structure. A capacitor consisting of a ferroelectric and a linear dielectric film represents a voltage divider. If an external voltage (V_G) is applied, the voltage across the ferroelectric layer (V_{FE}) is given by:

$$V_{FE} = \frac{V_G}{\dfrac{\varepsilon_{FE}d_{IL}}{\varepsilon_{IL}d_{FE}} + 1}, \tag{2.4}$$

where ε_{FE} and ε_{IL} are the relative dielectric constants of the ferroelectric and linear dielectric layer, respectively, and d_{FE}, d_{IL} are corresponding layer thicknesses. Equation (2.4) shows clearly that introduction of a buffer layer reduces the voltage drop across the ferroelectric film. A depolarisation field (E_{DEP}) appears as a result of incomplete compensation of the ferroelectric polarisation (P) at the electrode side. Insulating layers embedded between the electrode and the ferroelectric enhance this depolarisation field [93], [109]. Since E_{DEP} is directed opposite to the ferroelectric polarisation of the ferroelectric film, it leads to a polarisation reduction and deteriorates retention. In the simplified case of a ferroelectric capacitor with an insulating layer present at one of the electrodes E_{DEP} can be calculated as following [110]:

$$E_{DEP} = -\frac{P}{\varepsilon_0 \varepsilon_{FE}\left(1 + \dfrac{\varepsilon_{IL}d_{FE}}{\varepsilon_{FE}d_{IL}}\right)}. \tag{2.5}$$

The voltage losses in the buffer layer as well the depolarisation field can be diminished by using buffer materials with very high dielectric constants or by increasing the thickness of the ferroelectric layer. For ferroelectrics like PZT and SBT that exhibiting ε-values about 200 – 300, it is, however, difficult to find buffer materials with sufficiently high dielectric constants.

Therefore, a combination of both approaches with high-*k* HfO_2-based buffer layers ($\varepsilon_{IL} \sim 20$) and thick ferroelectric films (200 – 500 nm) is currently adopted [111]. The physical height of the contemporary FeFET gate stack imposes constraint on its lateral down-scaling with a limit at the 50 nm node [13]. The most aggressively scaled FeFET devices based on the perovskite ferroelectrics reported in the literature has achieved a gate length of 260 nm up to now [112]. Thus, despite the promising scaling potential of the 1T ferroelectric memories, their scaling is limited in practice by the conventional ferroelectric materials.

Another possible design of a 1T ferroelectric cell is the metal-ferroelectric-metal-insulator (MFMIS) transistor [113], [114] (Figure 2.12 right). Its gate stack is more complex in comparison to the MFIS cell with an additional metal layer embedded between the ferroelectric and buffer film. This additional metal layer serves as a floating electrode. The MFMIS stack structure enables the processing of the MFM and MIS capacitors to be decoupled. As a result, an improved resistivity against detrimental interdiffusion and a higher quality semiconductor interface can be obtained. The voltage drop across the buffer layer and the depolarisation field can be reduced by adjusting the ratio between the areas of the MIS and MFM capacitors (A_I/A_F), producing lower operation voltages and better retention characteristics [115]. However, taking into account the difference between the dielectric constants of the typical ferroelectric materials and known buffer layers, the A_I/A_F ratio must be at least 5 – 10 [115]. This complicates the fabrication process and also limits the scaling possibilities. Moreover, a single leakage path in the buffer layer short-circuits the floating electrode with the semiconductor surface leading to an immediate degradation of the memory function. On the contrary, in the MFIS cells, the weak spots in the buffer layer induce only local compensation of the polarisation charge and are less critical.

The existing FeFET prototypes demonstrate characteristics superior to the modern FG technology: high endurance resistivity up to 10^{12} program/erase cycles, 3 – 4 times lower operation voltages of 4 – 6 V and significantly shorter writing times in the range of nanoseconds [105], [111], [112]. Nevertheless, ferroelectric transistors have been left out of consideration as candidates for future memory applications due to the scaling issues and the incompatibility with CMOS processing associated with conventional ferroelectric materials. Therefore, there is a demand for new ferroelectric materials with characteristics that can enable the current limitations to be overcome. The required properties include a low dielectric constant, low remanent polarisation, high coercive field and compatibility with CMOS technology [116]. A promising candidate is the recently discovered HfO_2-based ferroelectrics, which will be discussed in detail in the next section.

2.3.3   Ferroelectricity in HfO$_2$

The polymorphs of HfO$_2$ most commonly reported in the literature are the monoclinic (P2$_1$/c), tetragonal (P4$_2$/nmc) and cubic (Fm3m) phases [117]. The monoclinic phase is the most thermodynamically stable in the bulk under ambient conditions. Elevated temperatures are required to induce a phase transition to the tetragonal and cubic phases at atmospheric pressure (1720 °C for the monoclinic-to-tetragonal and 2600 °C for the tetragonal-to-cubic transitions) [117]. Under high-pressure conditions two additional polymorphs of HfO$_2$, both having orthorhombic symmetries (Pbca and Pmnb), were observed [118]. All crystal phases known for bulk HfO$_2$ are centrosymmetric and, thus, cannot possess ferroelectric behaviour.

The properties of thin polycrystalline HfO$_2$ films can, however, differ significantly from their bulk counterparts. The contribution from the surface energy becomes comparable to the volume energy, especially in very thin layers with nanometre-size crystallites (1 – 5 nm), which, affects their physical properties [120], [121]. For instance, in thin films, the high-temperature polymorphs can be stabilised at ambient conditions by the incorporation of cation dopants [122], [123] – [126], the growth of ultra-thin films [127] or crystallization in the presence of capping layers [128]. The ionic radius of the impurities determines whether the tetragonal or the cubic phase becomes stable [122]. For example, doping of HfO$_2$ with silicon was reported to facilitate the formation of the tetragonal phase [123], [120], [129]. Furthermore, the stability of the amorphous phase in thin films is extended to higher temperatures in comparison to bulk HfO$_2$. The onset of the crystallization can be tuned by both the film thickness and the doping concentration [127], [120]. Thin HfO$_2$ films can exhibit electric properties that are different from those of bulk material, ferroelectric or antiferroelectric-like, instead of paraelectric. The effect of ferroelectricity in thin HfO$_2$ films was first discovered with Si-doping [18] and subsequently also demonstrated for several other tetravalent and trivalent dopants such as Zr [19], Y [20], Al [130] and Gd [131]. The origin of the ferroelectric behaviour in HfO$_2$ films is not yet completely understood. The ferroelectric properties emerge at a certain film composition that corresponds to the phase boundary between monoclinic and high-temperature phase (cubic or tetragonal), when film crystallisation occurs in the presence of mechanical confinement in the form of a capping layer (e.g. TiN electrode) [18], [119]. The amount of doping required depends on the dopant used and varies from several cation percent for Si, Al, Y, Gd (3 – 9 cat%) [18], [20], [130], [131] up to a fraction equal to Hf content as in case of Zr (30 – 50 cat%) [132]. The ferroelectric films can be obtained using atomic layer deposition (ALD) [18], [20] as well as physical vapour deposition (PVD) [133] techniques. The ferroelectric behaviour is proposed to originate from the non-centrosymmetric orthorhombic phase of the space group Pbc2$_1$ [18], [22], which emerges from the metastable tetragonal phase P4$_2$/nmc

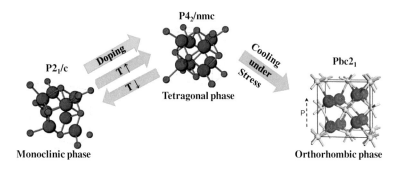

Figure 2.13 Relationship between the HfO_2 polymorphs according to [119].

exposed to the mechanical stress provided by the capping layer. This orthorhombic phase has already been reported to form during martensitic transformation from the metastable tetragonal phase in magnesia-partially-stabilized zirconia ceramics [134], [135]. The latter should normally transform into the monoclinic phase. However, as argued by Kisi et al. in [136], this phase transition is inhibited in the presence of internal lattice strains. The main reason is, that the tetragonal-to-monoclinic phase transition is accompanied by a volume expansion of about 3.5% [117] and requires shearing and twinning of the unit cell. This becomes unfavourable in the presence of internal lattice strain. As a result, a tetragonal-to-orthorhombic transition with less volume expansion and shearless unit cell transformation takes place instead. Due to the similarity of the crystalline structures of ZrO_2 and HfO_2, the occurrence of the orthorhombic $Pbc2_1$ phase can be also expected in HfO_2. Since this phase belongs to the non-centrosymmetric space group, it can potentially exhibit ferroelectric properties. Figure 2.13 shows the relationship between the monoclinic, tetragonal and ferroelectric orthorhombic HfO_2 phases.

The ferroelectric properties of thin HfO_2 films are of a particular interest for ferroelectric memories, since implementation of this material has the potential to overcome the limitations associated with conventional ferroelectrics. One of the main advantages of HfO_2 is its full compatibility with the conventional CMOS process. Since 2007, HfO_2-based materials have been introduced into CMOS technology as a high-k replacement to SiO_2 dielectrics in order to enable device downscaling below the 45 nm node [137] and are already established as reliable gate dielectrics in high-k metal gate technology [138], [139]. Moreover, HfO_2-based ferroelectrics provide a better scaling potential in comparison to the conventional ferroelectrics. Stable ferroelectric properties at film thicknesses in the nanometre range (5 – 30 nm) [22], [23] allow for a drastic reduction of the gate stack height, providing gate stack aspect ratios more suitable for scaling. The better scaling potential of HfO_2-based ferroelectrics is also assisted by a significantly lower dielectric constant of ~25 (PZT or SBT ~200 – 300) and a substantially higher coercive field strength E_C of ~1 MV/cm (PZT or SBT ~50 mV/cm). At reduced ferroelectric thickness the former enables high depolarization fields to be avoided whereas the latter compensates for the memory window loss. The possibility to use ALD technique for film fabrication ensures high-quality films with excellent conformity as well as precise thickness control in the nanometre range. FeFETs were fabricated at a state-of-the-art 28 nm technology node utilizing ferroelectric Si:HfO$_2$ [24], which finally closed the scaling gap between the ferroelectric and CMOS transistors. Furthermore, ferroelectric HfO$_2$ is suitable for integration into 3D structures [140]. This 3D integration provides potential for continuing the FeFET scaling and fabricating it in non-planar configurations, such as the FinFET and 3D array architectures that are projected for the technology nodes below 20 nm [4]. Thus, the HfO$_2$-based ferroelectrics can be expected to establish FeFET devices as a competitive concept for future memory applications and to enable their industrial implementation.

2.3.4 Traps in HfO$_2$

The implementation of HfO$_2$-based materials in CMOS technology has revealed their main deficiency, namely high intrinsic defect densities (10^{12} – 10^{14} cm^{-2}) [141], [142]. The reasons for the high defect concentration in HfO$_2$ in comparison to SiO$_2$ is the ionic type of bonding that includes electrons from the d-shells [143], [144] in combination with a high coordination number. This complicates the relaxation and re-bonding of the oxide network as it is the case for SiO$_2$, where covalent bonding and low coordination number enable the self-curing of defects [144]. These intrinsic defects can serve as electron [144], [141] and/or hole traps [145], [146]. As a result transistors including HfO$_2$-based dielectrics suffer from mobility degradation [147], [148], V_{TH} instability [141], [149] and reliability issues [150] – [152] such as negative and positive bias temperature instability as well as enhanced stress-induced leakage current. Fast electron trapping [141], [149], [153] and detrapping processes [154] in the sub-microsecond time ranges were detected using single-pulse measurement technique (chapter 3.3.2), providing evidence for the existence of shallow electron traps. Tunneling was identified as the main mechanism for fast trap charging and discharging [142], [154]. Attempts have been made to use trapping potential of HfO$_2$-based dielectrics by implementing them as storage layer into charge-trapping memories [155], [156]. These devices demonstrated faster programming speed in comparison to the standard SONOS cells owing to shallower traps at the expense, however, of the retention properties [157], [158].

Oxygen vacancies [144], [159] – [161] and oxygen interstitial atoms [144], [161], [162] are considered to be the main origin of traps in HfO$_2$-based materials. On the other hand defects including Hf ions, are expected to be energetically unfavourable [144]. It has been suggested that oxygen vacancies introduce shallow (E_T = 0.3 – 1.0 eV) as well as deep electron trap states (E_T >1.5 eV) into the HfO_2 band gap depending on their charging state [142], [152], [159], [161], [163]. The assignment of the trap energy levels to the individual charged states in the literature is rather contradictory at present. Moreover, negative U properties were predicted for oxygen vacancies due to the strong electron-lattice interaction [145], [159], [164]. As a result, trapping of two electrons/holes must be more favourable than trapping of single electron/hole, which would enhance trapping of electrons as well as holes.

3 Characterisation methods

This chapter gives an overview of characterisation methods utilised in this work. These include electrical measurements identifying the memory cell performance as well as physical characterisation methods aiming for material aspects of the memory cells. Electrical characterisation tests common for all memory types (read, program/erase, retention, endurance) as well as ferroelectric memory specific tests (PUND, polarisation-voltage measurements) are described. The basic principles of applied microstructural analyses (grazing incidence x-ray diffraction (GI-XRD), x-ray photoelectron spectroscopy (XPS), transmission electron microscopy (TEM) and piezoresponse force microscopy (PFM)) are elucidated. Furthermore, details of techniques used for characterisation of trapping behaviour in dielectric films, such as charge-pumping and single-pulse methods, are introduced.

3.1 Memory characterisation tests

3.1.1 Read/Sensing

Read or sensing operation is aimed to identify one of the possible states of a memory cell. In the simplest case there will be only two possible states – "ON" and "OFF" state. "ON" state is assigned to a high conductivity of the transistor channel (i.e. low V_{TH} value), while "OFF" state – to a low conductivity of the transistor channel (i.e. high V_{TH} value) (Figure 3.1). Both existing read approaches exploit shift in the I_D-V_G characteristic of a memory transistor depending on its state. In the first approach the V_{TH} value is monitored to determine the memory state (Figure 3.1 (a)), whereas in the second one drain current is measured at a constant gate voltage ($V_{G\ SENSE}$) for the same purpose (Figure 3.1 (b)). The first concept was used for read operation in this work. The I_D-V_G characteristics were recorded at a constant drain voltage of 100 mV with the source and bulk kept at 0 V. The V_{TH} value was extracted from the resulting I_D-V_G curves using a constant drain current criterion [26], [165]. V_{TH} value is defined as a gate voltage corresponding to a specified value of drain current (I_{TH}). This I_{TH} depends on the device geometry and can be calculated according to [165] as:

$$I_{TH} = 10^{-7} \cdot \left(\frac{W}{L} \right) [A] , \qquad (3.1)$$

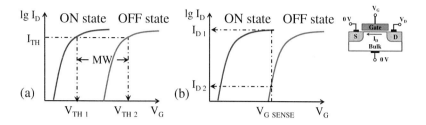

Figure 3.1 Read operation approaches for a transistor memory cell: (a) constant current method and (b) constant voltage method.

with W – transistor gate width and L – gate length. This current criterion was found to lie in the transition region between the exponential and linear I_D dependence on V_G, i.e. between the subthreshold and linear transistor operation regimes. Therefore, V_{TH} defined as $V_G(I_{TH})$ corresponded well to its physical meaning as a voltage required to build a transistor channel (see chapter 2.1.1). In order to minimise disturb during sensing operation, I_D-V_G measurements were restricted mainly to the subthreshold region and interrupted as soon as monitored I_D value exceeded defined I_{TH} criterion. All current-voltage measurements shown in this work were performed with Keithley's SCS-4200 analyser.

3.1.2 Program/Erase characteristics

During write operation (program or erase) one of the possible memory states is written into the memory cell. In order to determine writing speed and required voltage amplitudes program/erase characteristics are recorded by applying voltage pulses with varying lengths and amplitudes to the gate, while other device terminals are biased at 0 V. The induced V_{TH} shift is subsequently measured. During program operation "OFF" state with high V_{TH} value is

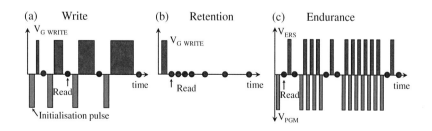

Figure 3.2 Measurement sequence during (a) write operation (program or erase), (b) retention and (c) endurance tests.

set, while during erase operation – "ON" state with low V_{TH} value is restored. Thus, in case of ferroelectric memories the program operation is performed with negative gate voltages, while during erase operation positive gate voltages are required. For charge trapping memories it is vice-versa. Figure 3.2 (a) schematically shows as an example a test sequence used for obtaining erase characteristic of a ferroelectric transistor. Write pulses with constant amplitude and logarithmically increasing length are applied to the gate and the memory state is read out in between as described in section 3.1.1. Prior to each writing pulse (program or erase) the identical memory state (completely erased or programmed) is re-established by applying an initialization pulse of the opposite polarity in respect to the writing pulse. By performing the same procedure for write pulses with different amplitudes, the most efficient operation conditions, enabling high write speed at sufficiently low voltages, can be identified.

3.1.3 Retention test

Retention is one of the main characteristics of the non-volatile memories, characterising their ability to retain the stored information over a long period of time without power supply. Typical requirement is 10 years data retention before the first erroneous readout [1]. This property is verified during the retention test (Figure 3.2 (b)). Initial gate pulse is applied to write "ON" or "OFF" memory state; the read out is subsequently performed with logarithmically increasing time delays. All device contacts are left idle during the waiting time. This technique emulates the case of unpowered storage and is more close to the real memory application conditions. The potential of 10 years storage can be verified either by extrapolating the experimental data or by performing temperature or bias acceleration tests [1].

3.1.4 Endurance test

Endurance test characterises the ability of the device to withstand electrical stress during continuous program and erase operations [1]. Figure 3.2 (c) depicts the standard measurement sequence. Pulses of alternating polarity, emulating program and erase operations, are applied to the device gate with other electrodes grounded. After certain number of stress pulses the capability to switch the memory cell into the "ON" and "OFF" state is tested by reading out the cell's V_{TH} values after one erase and one program pulse. Endurance characteristic provides the maximal number of program-erase cycles, to which the device can be exposed with individual memory states remaining distinguishable. So, for example, the current endurance specification of Flash memories is $10^4 - 10^5$ cycles [1].

3.2   Ferroelectric memory specific characterisation tests

The specific tests for ferroelectric materials include polarisation-voltage and PUND measurements, which can be performed using capacitor structures as well as transistor structures, where source, bulk and drain terminals are set at the same potential. Small-signal capacitance-voltage measurements are used to verify the ferroelectric/antiferroelectric film properties in addition to the polarisation test. The microscopic study of the domain structure and their switching kinetics is performed using PFM (piezoresponse force microscopy) technique.

3.2.1   Polarisation-voltage measurement

Polarisation measurements are used to determine the characteristic parameters of ferroelectric materials – the remanent polarisation (P_R) and the coercive field strength (E_C). Here, the total charge of the ferroelectric capacitors is monitored as a function of applied voltage. The standard approach to obtain polarisation curves is to use the Sawyer-Tower circuit [166]. A ferroelectric capacitor (C_F) and a reference capacitor (C_R) are connected in series to a triangular AC voltage source (Figure 3.3 (a)). The voltage drop at the reference capacitor (V_R), proportional to the polarisation charge on each capacitor, is measured and plotted over the total applied voltage giving a hysteresis curve. The polarisation charge is calculated as $P = C_R \cdot V_R / A_F$, with A_F being the area of the ferroelectric capacitor. Parasitic cabling capacitance and parasitic voltage drop across the reference capacitor are the shortcomings of the Sawyer-Tower circuit, which impair the measurement results. Moreover, hysteresis loops obtained with this approach cannot serve as unambiguous proof of the material's ferroelectric properties [68], [69]. Hysteresis behaviour can also arise from several experimental artefacts, e.g. surface polarisation, trapping-detrapping phenomena at Schottky-like electrodes [70] or leakage currents [71]. In order to ascertain the true behaviour of the studied sample the current response should be monitored as a function of applied AC voltage. In case of a ferroelectric sample two characteristic current peaks, corresponding to the domain switching at the coercive voltages (not at maximum voltage), should appear, as it is seen in Figure 3.3 (c). Integration of the transient current over measurement time provides polarisation values. These give the desired polarisation characteristic (Figure 3.3 (d)), if plotted versus excitation voltage. The values of the remanent polarisation (P_R) and the coercive voltage (V_C) can be extracted from cross-sections with polarisation- and voltage-axis, respectively. If the current maxima coincide with the maxima of the excitation signal, the leakage currents underlie the experimentally observed hysteretic behaviour. The virtual ground circuit (Figure 3.3 (b)) is a current-based approach utilised for polarization measurements. Here, the current response of a ferroelectric capacitor exposed to alternating voltage is monitored using a feedback resistor across an operational amplifier [167]. The non-inverting input is connected to ground. The signal from the ferroelectric capacitor is fed at the inverting input, which is simultaneously connected to the output via the feedback resistance (R). The output voltage adjusts automatically to achieve the equilibrium state with equal voltages at both inputs. The inverting input must be pulled to the ground level, so that the output voltage yields $V = -I \cdot R$. In the virtual ground circuit the parasitic effects of cabling capacitance and voltage losses on the reference capacitor, known for the Sawyer-Tower set-up, are eliminated.

In this work the polarisation-voltage measurements were performed using the aixACCT TF Analyser 3000 with implemented virtual ground measurement approach. All polarisation measurements were performed at 1 kHz frequency unless mentioned otherwise. Test sequence (Figure 3.3 (e)) includes 4 triangular pulses, which are applied consecutively with a time delay of 1 s [168]. This test sequence enables to extract not only values of dynamic remanent polarisation (P_R) at zero gate voltage but also values of relaxed remanent polarisation (P_{R_rel})

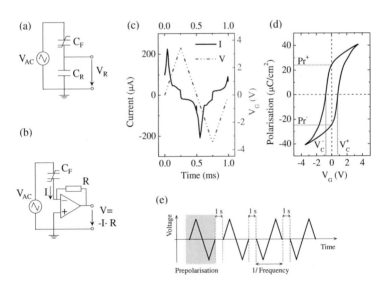

Figure 3.3 Polarisation-voltage measurements. Equivalent test circuits: (a) Sawyer-Tower circuit and (b) Virtual Ground circuit; (c) Transient current response of a ferroelectric capacitor under triangular excitation gate voltage during polarisation measurements and (d) resulting polarisation-voltage curve; (e) Test sequence for polarisation-voltage measurements utilised in this work.

remaining after 1 s delay for both polarisation states. The difference $\Delta P_R = P_R - P_{R\_rel}$ denotes the loss of remanent polarisation within 1 s and, thus, short-term retention properties of ferroelectric materials.

3.2.2 PUND measurement

PUND technique utilises a pulsed measurement approach to obtain polarisation characteristics. Here, a sequence of short square pulses instead of a continuous excitation signal is applied to the device. This pulsed characteristics are more relevant for the memory application, since they emulate the actual operation conditions of the semiconductor memory devices exposed commonly to high frequency pulses [169]. By means of the PUND measurements ferroelectric switching properties – switching time constants at different operation voltages – can be identified. The switching behaviour in the pulsed mode is different from that under the continuous excitation signal due to the frequency dependence of the coercive field [53]. As a result, the switching voltages required during pulsed operation are commonly higher than coercive voltages obtained from polarisation hysteresis loops (Figure 3.3 (d)). The typical PUND excitation signal consists of five consecutive pulses (inset of Figure 3.4 (a)): negative write pulse setting the defined initial polarisation state followed by two positive and two negative pulses (**P**ositive switching, **U**p non-switching, **N**egative switching, **D**own non-switching pulses). During the applied pulse train the current response is recorded. The area under the current transient corresponds to the polarisation charge. Figure 3.4 (b) depicts the case of switching with positive pulses. During the first positive pulse the measured current includes both switching and non-switching components;

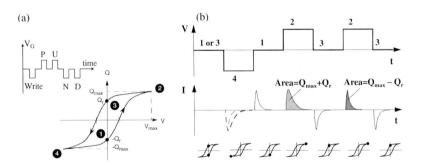

Figure 3.4 PUND measurement methodology [49]. (a) The pulsed polarisation hysteresis of a ferroelectric capacitor. Inset depicts typical PUND pulse sequence. (b) Wave train applied on the gate for measuring positive branch of the hysteresis loop, corresponding transient current response and polarisation states.

while during the second positive pulse only non-switching response is present. The subtraction of these current responses from each other gives the pure ferroelectric switching component, whereas additional contributions from dielectric polarisation, leakage currents and trapping are excluded. Integration of the current difference from the switching and non-switching pulses yields the twofold value of the pulsed remanent polarisation. PUND testing methodology, setup and measurement limitations are discussed in more detail in [169], [170]. In this work PUND measurements were carried out using aixACCT TF Analyser 3000.

3.2.3 Capacitance-voltage measurements

The small-signal capacitance-voltage characteristics were used to confirm the material properties derived from the polarisation measurements. As discussed in chapter 2.3.1 (see discussion to Figure 2.7) the form of a capacitance-voltage curve depends on the material class. During the capacitance-voltage measurements a small-signal differential capacitance is determined as a function of a gate bias. This differential capacitance is proportional to the displacement current induced by an AC signal of low amplitude (in this work 35 mV) applied to the gate superimposed with a bias voltage. Real capacitor structures include commonly parasitic components. The simplified equivalent circuit for a real capacitor can be represented by a capacitance (C) and a resistance (R) connected either in parallel or in series. In the commercial test systems the amplitude of the resulting displacement current and its phase shift to the excitation AC signal are measured. These can be converted into the R and C values with the help of translation equations, if the equivalent circuit is known. Serial equivalent circuit (Cs-Rs) is more appropriate in case of high serial resistance (contact issues, high substrate resistance), while in case of high leakage current, parallel equivalent circuit (Cp-Rp) should be chosen. Commonly the examined structure is more complicated than a two-element circuit. Thus, the decision on the proper measurement mode can be difficult. The dissipation factor – ratio between the real and the imaginary part of experimental impedance ($D = R/|X| = 2\pi f \cdot RC$) – can be used to validate the accuracy of the measurement. For $D < 0.01$ the phase shift between the displacement current and the excitation voltage signal converges to 90°, as in case of an ideal capacitor, and the capacitance measurement can be considered as reliable. The D-value and, thus, the contribution of parasitic resistances can be affected by changing the test frequency (f). High frequencies should be used in case of leaky capacitors to eliminate the impact of parasitic parallel resistance. The impact of high series resistance, on the contrary, is minimised at low test frequencies. The capacitance-voltage measurements presented in this work were performed with Agilent 4285A precision LCR Meter or multi-frequency impedance measurement card of a Keithley's SCS-4200 analyser.

3.2.4 Piezoresponse force microscopy

Piezoresponse force microscopy (PFM) provides a possibility for local imaging of ferroelectric domains at nanometre scale as well as for investigation of switching dynamics of single domains [171]. PFM was developed from the atomic force microscopy (AFM) technique [172] and makes use of its measurement set-up, which is schematically illustrated in Figure 3.5 (a). The surface of the sample is scanned with conductive probe, consisting of a flexible cantilever and a pyramidal shape tip. The lateral resolution of the system is determined by the radius of curvature of the tip apex, which is usually 10 – 20 nm. The measurements are performed in contact mode, where the probe tip is in direct contact with the investigated surface. The deflections of the probe cantilever are sensed by means of optical technique. A laser light from a solid state diode reflected off the cantilever back is collected by a photodiode detector. A displacement of cantilever results in one photodiode collecting more light than others, producing an output signal, which is fed into a differential amplifier. A quadrupole detector, containing four closely placed photodiodes, enables to distinguish between out-of-plane (vertical cantilever movement) and in-plane (torsional cantilever movement) surface deflections. The probe is mounted on a piezo scanner – a tube of piezoelectric ceramic. It regulates the probe vertical position by comparing signal from the photodetector with the specified set point value and adjusting the cantilever height and torsion until both signals match. This method allows keeping the contact force with the sample constant during the entire scanning process.

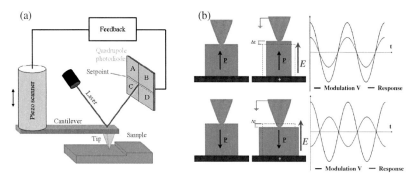

Figure 3.5 PFM operating principle [76]. (a) Schematic representation of a measurement set-up. (b) Piezoelectric response depending on the polarisation orientation of domains. If polarisation direction is aligned with applied electric field, in-phase piezoelectric response is observed. Anti-parallel orientation of domain polarisation and applied electric field results in 180° phase shift of piezoresponse to driving voltage.

The domain imaging principle is based on the converse piezoelectric effect, which consists in inducing sample deformation by external electric field. In order to detect low level piezoelectric response, which is typically close to the resolution limit of high precision AFM tools, AC excitation signal is used in combination with lock-in technique. The AC bias applied on the probe tip induces surface oscillatory deformation. The latter is transmitted to the cantilever, whose resulting movement is detected by a photodiode detector and converted into an oscillating voltage output signal. Lock-in amplifier compares this output signal with the reference signal having the frequency of the excitation signal and singles out the component at the reference frequency, while removing all signal components with other frequencies. In this way very small alternating signals with amplitudes thousand times lower than noise can be measured. The amplitude and the phase of the separated signal are then retrieved by the lock-in amplifier as a piezoresponse amplitude (*PRampl*) and piezoresponse phase (*PRphase*). The amplitude of the piezoresponse can be used to quantify the local piezoelectric constant. The phase of the piezoresponse gives, on the other hand, information on the orientation of domains' polarization relative to the external field E (Figure 3.5 (b)). If domain polarisation coincides with the direction of the external field, they will oscillate in-phase with the excitation AC signal. In this case the positive bias induces expansion of the domains, while negative bias – domain contraction. On the contrary, domains with polarization reversed to the external field will oscillate with 180° phase shift to the excitation AC signal. Here positive bias results in contraction of domains and negative bias in their expansion. In this way, by using PFM technique, domains with different polarization directions can be distinguished.

Besides visualization of the existing domain structure, PFM also enables manipulation of the domain polarization [173]. PFM scans can be executed either in reading or writing mode. During reading the amplitude of the applied bias is below coercive voltage, so that only piezoelectric response of domains is sensed without affecting their polarization direction. In a writing mode the sample is exposed to the voltages higher than characteristic coercive voltages, so that the area underneath the probe tip gets polarized during scanning process. Using probe as a mobile electrode domains of arbitrary shapes as well as domain arrays can be written [174], [175]. Furthermore, the microscopic domain switching mechanisms can be studied [176].

Piezoresponse hysteresis measurements, showing dependence of the piezoelectric response on the constant bias field, are used to analyse the local switching behaviour of ferroelectric materials [177]. By sweeping DC voltage from negative to positive values and back, the characteristic results in butterfly-shaped characteristics of piezoelectric constant (*PRampl*) [77], [56] and hysteretic curves of *PRphase* similar to the polarization hysteresis loops [173]. In-field hysteresis loops are obtained by recording the piezoresponse in the presence of constant electric field. Here, the AC piezoresponse measurement signal is superimposed with DC bias. The off-field piezoresponse loops arise from the measurements after individual DC bias is turned off, reflecting material local retention properties. The band excitation method, discussed in detail in [178] and [179], enables to eliminate the cross-talk between the film topography and the piezoelectric response.

PFM measurements on $Si:HfO_2$, studied in this work, were performed by Dominik Martin. AFM system in combination with a dual-phase SR830 DSP Lock-in Amplifier was used for this purpose. Piezoresponse images were acquired in ambient air. Domain poling experiments were carried out on free $Si:HfO_2$ surface. Hysteresis measurements were performed on the bonded pads with the purpose to minimise electrostatic effects: the DC voltage signal (applied to the pad) was decoupled from AC measurement signal (applied to the probe tip).

3.3 Trapping characterisation methods

3.3.1 Charge pumping

Charge pumping (CP) is a standard method used to perform in-depth analyses of the interface quality in the MOSFET devices. By means of CP technique the interface traps, located directly at the Si / gate dielectric interface, as well as the near-interface bulk traps, located within an insulator layer at depths of several nanometres from interface, can be characterised. Trap densities as low as 10^9 traps/eV/cm^2 can be determined reliably owing to a very high sensitivity of this method [180]. As the measurements are carried out directly on the MOSFET structures, direct correlation between processing conditions and interface quality can be obtained [180].

The basic measurement setup to perform charge pumping measurements, as introduced by Brugler and Jespers in [181], is shown in Figure 3.6 (a). The gate is connected to the pulse generator. The source and drain are tied together and connected to the ground or reversed biased with respect to the substrate. Applying periodical pulses with sufficiently high amplitude to the gate, DC current (referred as charge-pumping current) will be measured at the substrate and source/drain contacts. This charge-pumping current originates from the recombination process between minority and majority carries, which occurs via interface traps [180], [182]. If the pulsed signal with the maximum level above the threshold voltage (V_{TH}) and minimum value below the flatband voltage (V_{FB}) is applied to the gate, the semiconductor

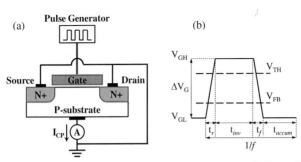

Figure 3.6 (a) Basic experimental setup for charge-pumping measurements. (b) Trapezoidal gate pulse, most commonly used, with the main parameters defined.

surface is continuously switched between inversion and accumulation, thus enriched alternately with minority and majority carriers. These are captured by the interface traps, where recombination process takes place. Thereby, minority charge carriers are pumped from source/drain junctions into the semiconductor substrate, giving rise to the CP current. The pulses most commonly used during CP measurements exhibit a trapezoidal form (Figure 3.6 (b)). It is defined by the following parameters: the highest signal level V_{GH}, the lowest signal level V_{GL}, pulse amplitude ($\Delta V_G = V_{GH} - V_{GL}$), rise time ($t_r$), fall time ($t_f$), time in accumulation (t_{accum}) and inversion (t_{inv}), pulse period ($1/f$), which is reciprocal to the signal frequency (f). Additional current components besides recombination current may arise from the leakage currents through the gate dielectric or geometric current [180]. Since these current components contain no information about interface traps, they must be eliminated during experiment. The geometric current originates from recombination of mobile inversion electrons, which do not have enough time to flow back to the source/drain regions, while substrate is switched from inversion into accumulation [181]. By using test structures with channel length (L) << channel widths (W) [183] and applying pulses with rise/fall times longer than 100 ns [182], the impact of geometric current can be almost completely eliminated. Contribution from leakage current reduces at high test frequencies.

Since the interface traps serve as recombination centres, the charge pumping signal is directly linked to the trap parameters, such as density of trap states, capture cross sections, energetic distribution within the band gap as well as spatial distribution along the channel. A quantitative model of the charge pumping current was developed by Groeseneken et al. [182]. The expression for the net charge pumping current (I_{CP}) is given by [182]:

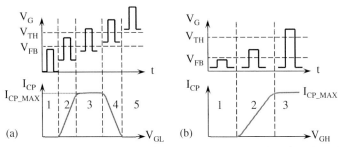

Figure 3.7 Schematic illustration of the two main CP measurement approaches after [180]: (a) variable base level test and (b) variable amplitude test.

$$I_{CP} = 2qkTfD_{it}A_{eff} \ln \left(v_{th} n_i \sqrt{\sigma_n \sigma_p} \frac{|V_{TH} - V_{FB}|}{\Delta V_G} \sqrt{t_r t_f} \right), \tag{3.2}$$

where D_{it} is the mean energetic density of surface states [traps/eV·cm^2], A_{eff} – the effective area of the channel under the gate electrode, v_{th} – the thermal velocity of the carriers, n_i – the intrinsic carrier concentration, σ_n/σ_p – capture cross sections for electrons/holes. As can be seen from (3.2) I_{CP} is directly proportional to the interface trap density, signal frequency and device area.

Two main approaches for performing the charge pumping measurements are: (1) variable base level test (Figure 3.7 (a)), originally proposed by Elliot [184], and (2) variable amplitude test (Figure 3.7 (b)), introduced by Brugler and Jespers [181]. In the variable base level test the amplitude of gate pulses (ΔV_G) is kept constant, while the lowest level (base level) V_{GL} is varied from inversion to accumulation. The I_{CP} versus V_{GL} characteristic reveals 5 regions (Figure 3.7 (b)). The rising edge of $I_{CP}(V_{GL})$ is located at V_{GL} approximately equal to ($V_{TH} - \Delta V_G$), whereas it's falling edge corresponds to $V_{GL} = V_{FB}$. The saturation value of the charge pumping current in the region 3 gives the average surface density of interface states N_{CP} [traps/cm^2], according to [184]:

$$I_{CP\_MAX} \approx fA_{eff}qN_{CP}, \tag{3.3}$$

averaged over the entire scanned energy range. During the variable amplitude test (Figure 3.7 (a)) the pulses of increasing amplitude are applied to the gate, while the lowest pulse level V_{GL} is kept constant and sets the semiconductor surface into strong accumulation (i.e. $V_{GL} < V_{FB}$). I_{CP} curve is plotted versus the highest signal level V_{GH}. When V_{GH} reaches the value of V_{FB} a strong increase in the I_{CP} signal is observed. It normally saturates after V_{GH} exceeds the V_{TH} value. This saturation level I_{CP_MAX} can again be used to estimate N_{CP} [traps/cm^2] according to (3.3). The non-saturated characteristics were suggested to originate from interaction of carriers with dielectric bulk traps located close to the interface and filled via tunnelling mechanism [185]. This phenomenon is especially pronounced for high-k dielectrics due to a high intrinsic concentration of bulk traps [186], [187], [188]. For the stacked gate dielectrics it is still under discussion, which traps are able to contribute to the charge pumping signal. Some authors argue, that only traps located within the interface layer and originate from the interaction with high-k material, can be sensed during the charge pumping experiment [189], [190], [191]. Others [192] claim, that bulk traps of the high-k materials can also contribute to the measured current. The contribution from bulk traps increases at low test frequencies, when charge carries have sufficient time to tunnel to the traps and back during inversion and accumulation time periods, respectively. It results in an increased charge pumping current as observed in [193], [194].

The mean energetic density of surface states D_{it} [traps/eV·cm^2] and geometric mean value of the capture cross section $\tilde{\sigma} = \sqrt{\sigma_n \sigma_p}$ can be determined from frequency dependent measurements under the assumption, that only interface traps contribute to the CP current, whereas the contribution from bulk traps is neglected. For trapezoidal pulses with rise and fall time changing with frequency as $t_r = t_f = \alpha / f$ (α – fraction of the pulse period), the pulse charge pumping current (3.2) becomes linearly dependent on the logarithm of frequency:

$$I_{CP} = 2qkTfD_{it}A_{eff}\left(\ln\left(v_{th} n_i \sqrt{\sigma_n \sigma_p} \frac{|V_{TH} - V_{FB}|}{\Delta V_G} \alpha \right) - \ln\left(f \right) \right). \tag{3.4}$$

According to (3.4) one obtains $\bar{D}_{it}$ from the slope of the $I_{CP}\left(\ln\left(f\right)\right)$ curve, while the intercept with the x-axis at $\ln\left(f_0\right)$ gives the $\tilde{\sigma}$ value:

$$D_{it} = \frac{1}{2qkTA_{eff}} \cdot \frac{d(I_{CP}/f)}{d\left(\ln f\right)}, \tag{3.5}$$

$$\tilde{\sigma} = \sqrt{\sigma_n \sigma_p} = \frac{\Delta V_G f_0}{v_{th} n_i \alpha |V_{TH} - V_{FB}|}. \tag{3.6}$$

3.3.2 Single-pulse I_D-V_G

Single-pulse I_D-V_G measurements are widely used to study the trapping and detrapping behaviour of high-k gate dielectrics [195], [196], [141]. The main advantage of this technique is its ability to capture the fast transient nature of the trapping and detrapping processes. Time resolution in nanosecond time range can be provided [197]. The testing is performed on transistor structures. This enables a direct correlation between fabrication conditions and trapping characteristics of the entire gate stack. The stress pulse of certain amplitude and width is applied to the gate, while the drain current is simultaneously monitored (Figure 3.8 (a)). The latter can be translated into I_D-V_G characteristics for both rising and falling pulse edges (Figure 3.8 (b)). From the value and the sign of the V_{TH} shift between these I_D-V_G curves conclusions about trapping/detrapping behaviour of the gate stack can be made. The main advantage of the single-pulse technique is that the time delay between stressing and sensing is practically eliminated. Therefore, the complete amount of trapped/detrapped charges can be captured. Figure 3.8 shows an example of the single-pulse measurement performed on a transistor with a gate dielectric stack consisting of 1.2 nm SiON interfacial layer and 10 nm Si:HfO$_2$. A positive shift between I_D-V_G curves, recorded at the rising and falling gate pulse edges (Figure 3.8 (b)), as well as the drain current degradation at the pulse plateau (Figure 3.8 (a)) indicate strong electron trapping into the high-k dielectric.

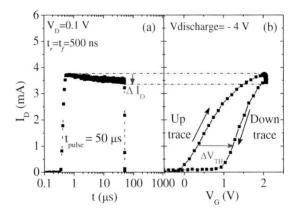

Figure 3.8 (a) Single-pulse characteristics measured on a transistor with a gate dielectric consisting of 1.2 nm SiON and 10 nm Si:HfO$_2$. The electron trapping can be identified either from (a) the shift between I_D-V_G curves measured at the up and down traces or (b) I_D degradation over pulse time.

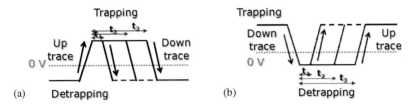

Figure 3.9 Gate pulse sequence for studying of (a) trapping and (b) detrapping kinetics.

For qualitative characterisation of trapping the value of the V_{TH} shift is usually used. The V_{TH} shift can be extracted either from the hysteresis plot of the I_D-V_G curves or calculated from the time dependent drain current drop ΔI_D at the pulse plateau [198]. According to [198] ΔV_{TH} values obtained with the first approach are more accurate, whereas the second approach gives underestimated ΔV_{TH} values. In the hysteresis approach ΔV_{TH} has to be determined in the range of drain currents, where I_D-V_G characteristics experience parallel shifting and are not affected by trapping or mobility degradation. The kinetics of trapping and detrapping processes can be studied by applying to the gate pulse sequences illustrated in Figure 3.9 (a) and (b), respectively. The trapping behaviour is examined by using pulses with positive amplitudes and varying widths. The pulses typically start at negative voltages to ensure initial complete discharging of traps. During the detrapping studies, on the contrary, pulses with negative amplitudes starting at positive voltages are applied. The start at positive voltages provides initial charging of traps. Figure 3.10 (a) shows the schematic diagram of the setup commonly used for single-pulse I_D-V_G measurements. Here a transistor is connected as in case of an inverter circuit with a load resistance R_{LOAD}. For each measurement a single pulse is applied to the gate of the transistor using a pulse generator, while its drain is simultaneously biased at a certain voltage. The source and bulk are grounded. The voltages at the gate and drain are simultaneously measured using an oscilloscope and then converted into current-voltage (I_D-V_G) characteristics. In this work a different setup, illustrated in Figure 3.10 (b), was implemented. Instead of external pulse generator for pulsing and oscilloscope for sensing, a program measurement unit (4225 PMU) of the Keithley´s SCS-4200 analyser together with two remote pulse amplifier/switch units (4225-RPM) were used. The PMU with two channels provides the capability of simultaneous high-speed voltage sourcing (time resolution 20 ns) and voltage and current measuring at rates of up to 200 Megasamples/s for each channel. The RPM units extend the accessible current range down to 100 nA with resolution less than 200 pA. By placing the RPM units in the immediate vicinity of the test device the parasitic cabling effects can be minimised. Thus, this equipment allows direct measuring of the drain current transients in a wide time range (pulse width from 20 ns up to 1 s) with high sensitivity. Moreover, the need of the drain current normalisation is eliminated

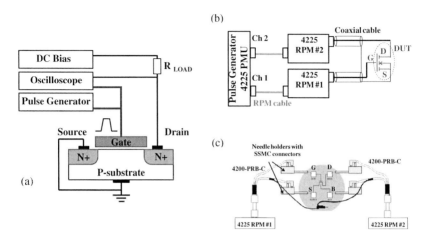

Figure 3.10 Schematic diagram of the setup for single-pulse I_D-V_G measurements: (a) commonly used setup, (b) experimental setup used in this work, (c) connections of the DUT with PRB-C adapter cables.

due to the constant drain voltage in contrast to the typical measurement approach (Figure 3.10 (a)) [141]. The RPMs are connected to the transistor test structure in the way shown in Figure 3.10 (c). With this measurement setup the pulse transition times could be set to 500 ns, whereas pulse widths were varied between 0.1 and 100 µs.

3.4   Microstructural analyses

3.4.1   Grazing incidence x-ray diffraction

X-ray diffraction (XRD) is a non-destructive method used to study atomic structure of a crystalline matter. Crystal symmetry, lattice parameters, lattice strain, qualitative and quantitative phase composition as well as preferred orientation (texture) of grains in polycrystalline materials can be obtained by means of XRD analyses [199], [200], [201]. This technique makes use of an x-ray wave interference on the periodic arrays of atoms in a crystal, which serve as a diffraction grating. This interaction becomes possible due to x-rays wavelength being of the same order of magnitude (1-100 Å) as the interatomic and interplanar distances in crystals. A constructive interference occurs only at certain diffraction angles determined by a specific crystal structure of a material, its crystal symmetry and lattice parameters. The condition of a constructive interference is described by the Bragg's law [200]:

$$2d_{hkl} \sin \theta_{hkl} = \lambda , \qquad (3.7)$$

where d_{hkl} is the spacing between the series of parallel lattice planes with Miller indices (h, k, l) responsible for a particular diffraction peak; θ_{hkl} is the diffraction angle – the angle, which the incident and the reflected beams make with the $(h\ k\ l)$ planes; and λ – the wavelength of the incident x-ray beam. The Miller indices define the crystallographic orientation of planes as well as spacing d between them in combination with the lattice parameters. Different crystallographic planes are characterised by a different interplanar spacing, which alter the θ angle of the corresponding diffraction peak. A diffraction pattern – a set of allowed diffraction angles and corresponding relative peak intensities, is unique for each crystalline matter. A large diffraction data base is available from the international centre for diffraction data (ICDD). It contains experimentally collected as well as theoretically calculated powder diffraction patterns for different materials and their polymorphs. These can be used as reference patterns for identification of the phase composition of the studied sample. Figure 3.11 (a) shows an experimental diffraction scan obtained on a 10 nm Si:HfO$_2$ film embedded between two TiN layers together with reference patterns of a tetragonal HfO$_2$ phase with a space group P4$_2$/nmc [202] and a cubic TiN (Fm3m) [203].

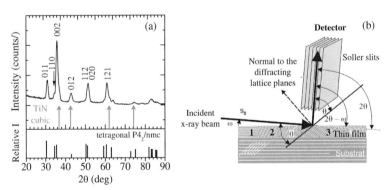

Figure 3.11 (a) Experimental diffraction data for a 10 nm thick polycrystalline Si:HfO$_2$ film embedded between TiN layers (top) and reference powder diffraction patterns for the tetragonal HfO$_2$ phase of a space group P4$_2$/nmc (bottom) [202] and the cubic TiN (Fm3m) [203]; (b) Schematics of a GI-XRD measurement approach [201].

Grazing incidence x-ray diffraction (GI-XRD) is an advanced XRD measurement technique, utilised for investigation of thin films with thicknesses in the range of a few nanometres [200]. The penetration depth of an x-ray beam in a standard symmetric geometry achieves $10 - 100$ µm depending on the absorption properties of the material. In this case the diffraction signal comes primarily from the substrate, whereas the contribution from the nanometre thick films is negligibly small. In the GI-XRD method (Figure 3.11 (b)) the incoming beam enters the sample under a very small angle of incidence (ω) of a few degrees or even less. This enables to increase the path travelled by the incoming beam within a studied thin film, so that the diffraction signal is magnified, while the signal coming from substrate is significantly reduced or even completely eliminated. The GI-XRD spectra are recorded with a constant incidence angle. The position of the x-ray source remains the same in respect to the sample, whereas the detector moving around the sample and collects the diffraction data from different diffraction angles.

GI-XRD was exploited in this work for studying the crystal structure and identification of phase composition in thin ferroelectric Si:HfO$_2$ films. The GI-XRD experimental diffraction data were collected using a Bruker D8 Advance diffractometer with Cu-Kα radiation ($\lambda = 1.5418$ Å) with an incidence angle of 0.5°.

3.4.2   X-ray photoelectron spectroscopy

X-ray photoelectron spectroscopy (XPS) is a non-destructive analytical method, which provides qualitative and quantitative information about chemical composition of solid materials within the topmost few nanometres ($1 - 10$ nm) of the surface [204]. This method is based on the photoemission of electrons from the surface of a sample exposed to monochromatic x-ray radiation. The kinetic energies and number of photoemitted electrons corresponding to different energy levels are being measured. The kinetic energy can be transformed into the electron binding energy using a law of energy conservation [204]. The set of binding energies is characteristic for each chemical element, whereas the number of emitted electrons (XPS peak intensities) is directly related to the amount of the element present within the studied sample.

By means of XPS technique the Si-content within the Si:HfO$_2$ thin films was analysed. XPS measurements were performed at Fraunhofer Center Nanoelectronic Technologies Dresden with a REVERA VeraFlex production metrology system using Al-Kα excitation and 141.2 eV pass-energy. The intensities of characteristic peaks for hafnium (Hf-4f) and silicon (Si-2p) were determined. The Si-content was subsequently obtained as cation fraction (cat%) by applying the tool specific calibration factors and weighting the areas of the respective XPS peaks.

3.4.3   Transmission electron microscopy

Transmission electron microscopy (TEM) is a microscopy technique enabling to study the sample's microstructure with sub-nanometre resolution. Utilisation of electrons with significantly smaller de Broglie wavelength (λ < 0.05 Å) instead of the visible light (λ in the range 3900 – 7000 Å) provides a considerably higher resolution capability of TEM in comparison with the light microscopy. By using TEM, both images and diffraction patterns of the specific specimen area can be obtained. These are formed by electrons transiting through a thin specimen. Due to interaction of transmitted electrons with the matter of the specimen, they contain information about its microstructure. Three main principles of the contrast formation in TEM images are distinguished [205]: mass-thickness contrast, diffraction contrast and phase contrast. The latter is utilised by a high-resolution TEM to image material structure on the atomic scale. More information about principle of image formation and imaging system of the high-resolution TEM can be found in [205], [206]. The samples studied with TEM should be thin enough (less than 100 nm) in order to obtain sufficient intensity of the transmitted electron beam. Therefore, specific specimen preparation from the bulk samples prior to actual TEM analyses is required. Mechanical thinning, electrochemical thinning or ion milling can be exploit for this purpose [204].

High-resolution TEM was used to study the microstructure of the polycrystalline Si:HfO$_2$ films. Thin specimens of different film regions were prepared for TEM analysis by means of focused ion beam (FIB) technique, a special type of ion milling, which utilises highly energetic gallium ions. High-resolution TEM measurements were performed by Dr. Thomas Gemming at Leibniz Institute of Solid State and Material Research Dresden.

4 Sample description

4.1 Metal-insulator-metal capacitors

The metal-insulator-metal (MIM) capacitors featured Pt/TiN/Si:HfO$_2$/TiN film stacks deposited on 300 mm silicon wafers. The process flow is schematically depicted in Figure 4.1. The fabrication of the capacitor stacks was performed at Fraunhofer Center Nanoelectronic Technologies Dresden (Fraunhofer CNT) by Johannes Müller, whereas subsequent annealing and structuring was carried out at NaMLab gGmbH by Andrew Graham. The hafnium oxide films were grown by means of water-based atomic layer deposition process at 300 °C using hafnium tetrachloride (HfCl$_4$) and silicon tetrachloride (SiCl$_4$) as metal precursors and N$_2$ as purge gas. The process was run in an ASM Pulsar® 3000 ALD tool, including a hot-wall reactor with a cross-flow design. In average, the film growth rate was 0.58 Å/cycle. The silicon content was tuned by the ratio of the SiCl$_4$ to HfCl$_4$ pulses. By varying the SiCl$_4$/(SiCl$_4$ + HfCl$_4$) ratio from 0 to 14 %, films with silicon content ranging between 0 and 8.5 cat% were obtained. The film composition corresponding to different deposition conditions was determined by means of XPS technique (for more details

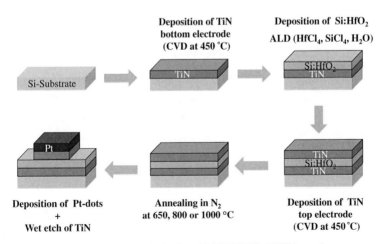

Figure 4.1 Process flow utilised for the fabrication of Pt/TiN/Si:HfO$_2$/TiN/Si capacitors.

see chapter 3.4.2). The same silicon doping series were run for three different film thicknesses of 9, 27 and 50 nm in order to study the impact of thickness on the properties of the $Si:HfO_2$ layers. The film thickness was controlled by changing the number of deposition cycles and later confirmed by spectral ellipsometry. The 10 nm thick TiN layers served as top and bottom electrodes. These were grown in a batch furnace using a pulsed chemical vapour deposition (CVD) process at $450^{\circ}C$ with $TiCl_4$ and NH_3 as precursors and N_2 as a purge gas. Additional NH_3 flushing was used in order to reduce chlorine contamination. For TiN films an average growth rate of 0.4 nm/cycle was detected. Following the formation of the $TiN/Si:HfO_2/TiN/Si$ stack, the 300 mm wafers were cut into 40 mm × 40 mm chips. These were subsequently exposed to three different annealing conditions: $650^{\circ}C$ for 20 s, $800^{\circ}C$ for 20 s or $1000^{\circ}C$ for 1 s. In all cases the annealing was performed in a nitrogen atmosphere in an AST rapid thermal processing (RTP) oven. The anneal at $1000^{\circ}C$ for 1 s was aimed to emulate the dopant activation anneal step, standard for manufacturing of CMOS devices. In the next step, the microstructure of the $Si:HfO_2$ film was studied using GI-XRD (chapter 3.4.1) and transmission electron microscopy (TEM) (chapter 3.4.3). As a final step single capacitor structures were fabricated for electrical characterisation. For this purpose the 40 × 40 mm chips were coated with 50 nm Pt by electron beam evaporation using a shadow mask to define dot arrays. Afterwards, these Pt dots were used as a mask to wet-etch the 10 nm TiN layer into individual electrode areas with an ammonia (0.5%) and hydrogen peroxide (1.2%) solution at 50°C for 5 minutes.

4.2   Ferroelectric field effect transistors

$Si:HfO_2$-based metal-ferroelectric-insulator-semiconductor field effect transistors (MFIS-FETs) with a poly-$Si/TiN/Si:HfO_2$/interface oxide/Si gate stack were fabricated using state-of-the-art 28 nm high-*k* metal-gate technology [139] on 300 mm industrial manufacturing equipment at GLOBALFOUNDRIES Dresden Module One LLC & Co. KG. Only minor adjustments to the overall integration were needed, since HfO_2 is already used a standard high-*k* gate dielectric in the contemporary CMOS technology. Devices with varying gate length scaled down to 28 nm were manufactured and tested. Figure 4.2 shows a TEM image of 32 nm device together with an enlarged image of the ferroelectric gate stack. The interfacial layer embedded between the silicon substrate and the ferroelectric layer was SiON (silicon oxynitride) obtained by decoupled plasma nitridation of the chemical oxide. It exhibited a thickness of about 1.2 nm. After the formation of the interfacial layer, the ALD process for the 9 nm thick $Si:HfO_2$ film followed. The process parameters similar to those described for the MIM capacitors in chapter 4.1 were used to deposit $Si:HfO_2$ layers with three different compositions (3.7, 4.4 and 5.7 cat% of Si). Physical vapour deposition (PVD)

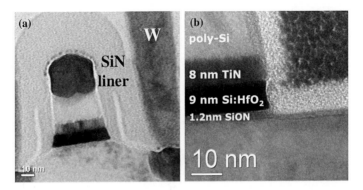

Figure 4.2 (a) TEM images of 32 nm Si:HfO$_2$-based MFIS-FET device and (b) an enlarged image of the ferroelectric gate stack.

at room temperature was used for growing 8 nm thick TiN layer aiming to avoid the shortcoming of the CVD process. Undesired partial crystallisation of the HfO$_2$ films was detected for the MIM structures during formation of the CVD-TiN top electrode as will be shown in chapter 5.2. In order to structure the Si:HfO$_2$ layers with physical thickness larger than that of a standard high-*k* gate dielectric (2 nm), a reactive ion etching process (RIE) at elevated temperatures was developed at Fraunhofer CNT and implemented during manufacturing of HfO$_2$-based MFIS-FETs. Sufficiently steep sidewalls with angles of about 85° could be obtained. The thermal budget for the complete gate stack reached a maximum temperature of 1050 °C at spike activation anneal, which resulted in a fully crystalline Si:HfO$_2$ ferroelectric.

5 Stabilisation of the ferroelectric properties in Si:HfO₂ thin films

The ferroelectric properties in HfO_2 were found to be induced by doping with various tetravalent and trivalent elements such as Si [18], Zr [19], Y [20], Al [130], and Gd [131]. The emphasis of this work lies on the properties of the Si-doped HfO_2 and its potential to be implemented into non-volatile one transistor memory cells. Si-doped HfO_2 ALD films grown from metal halide (here, chloride-based) precursors and H_2O were studied in contrast to the previous works [22], [207], where metal-organic precursors/O_3 process was utilised for film fabrication. Different precursor chemistry, oxidant and carbon contamination level as well as additional chlorine contamination can have an impact on the occurrence of ferroelectricity in the Si:HfO_2 system as well as on the electrical film properties. Therefore, material aspects of Si:HfO_2 thin films were studied first using planar capacitor structures (chapter 5) in order to get better insight into the ferroelectric properties and guidelines for subsequent transistor fabrication. The impacts of the silicon doping level (chapter 5.1), post-metallisation annealing (chapter 5.2) and film thickness (chapter 5.3) on the emergence of ferroelectricity will be discussed. Electrical characterisation combined with structural analyses enabled to find correlations between crystalline structure of films and their electrical properties. In addition, the true ferroelectric behaviour of Si:HfO_2 films were confirmed by means of piezoresponce force microscopy. All results shown in this chapter were carried out on MIM capacitor structures.

5.1 Impact of the silicon doping

5.1.1 Electrical characterisation

MIM capacitors containing Si:HfO$_2$ films with varying silicon concentration ranging from 0 (pure HfO$_2$) to 8.5 cat% were electrically characterised using polarisation-voltage (*P-V*) (chapter 3.2.1) and capacitance-voltage (*C-V*) measurements (chapter 3.2.3). Figure 5.1 shows the results for the 9 nm thick Si:HfO$_2$ films annealed at 1000 °C for 1 s. These annealing parameters are of particular significance, since they are similar to those used for dopant activation during transistor fabrication and, thus, emulate the thermal treatment of Si:HfO$_2$ layers within the FeFET stack. The film composition appeared to have a strong impact on the electrical properties. Transitions from paraelectric to ferroelectric and from ferroelectric to antiferroelectric-like behaviour were visible for increasing silicon content (Figure 5.1 (a)). The polarisation hysteresis loops alone cannot, however, serve as unambiguous proof of the ferroelectric or antiferroelectric properties [68], [69]. Therefore, the particular materials behaviour was additionally ascertained by the examination of the transient current response (Figure 5.1 (b)) and *C-V* characteristics (Figure 5.1 (c)). Examples of these characteristics and their correlation with material properties can be found in chapter 2.3.1. Pure HfO$_2$ exhibited paraelectric properties with a close to linear polarization response (Figure 5.1 (a)) and displacement current behaviour similar to a linear capacitor (Figure 5.1 (b)). Its *C-V* characteristics showed almost no dependence on the voltage and sweep direction. Doping of HfO$_2$ films with Si induced a quite different behaviour. Polarization hysteresis loops appeared for films with the lowest Si content of 4.4 cat%, indicating the presence of a ferroelectric phase. The remanent polarisation of 24 μC/cm^2 and the coercive field strength of about 0.9 MV/cm were extracted. Butterfly-shaped *C-V* curve characteristic for ferroelectric materials [72], [73] confirmed the assumption of truly ferroelectric behaviour. Moreover, the displacement current response exhibited two current peaks, associated with domain switching at the coercive voltages. A further increase in the Si content (5.6 – 8.5 cat% Si) induced a pinched hysteresis loops and double hysteresis loops similar to those of antiferroelectric materials. Furthermore, double-butterfly-shaped *C-V* curves and displacement current response with four switching peaks were detected for these film compositions. A similar dependence of the electrical behaviour on silicon concentration was also detected for other annealing conditions and film thicknesses (27 and 49 nm). Although the double hysteresis loops in combination with double-butterfly-shaped *C-V* curves are commonly attributed to the antiferroelectric materials, they are not necessarily an evidence of an antiferroelectric phase. Under certain conditions ferroelectric materials were reported to show a similar behaviour [69], [56]. The observed transition from the ferroelectric to antiferroelectric-like characteristics with increasing silicon doping can be explained in the context of several theories: defect dipoles or field-induced phase transition. Pinched and double hysteresis loops appear in conventional ferroelectric materials (e.g. PZT) as a result of doping with acceptor dopants. This effect is attributed to the increased concentration of defect dipoles, which

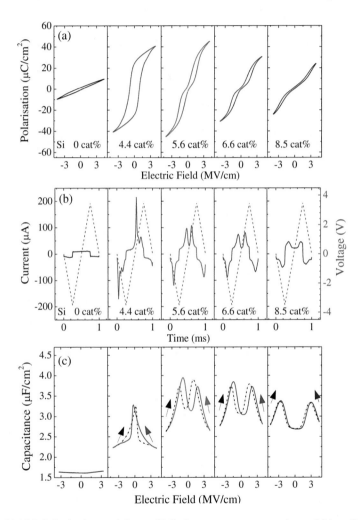

Figure 5.1 (a) Polarization hysteresis loops, (b) displacement current versus time and (c) small-signal capacitance-voltage characteristics obtained for TiN/Si:HfO$_2$/TiN capacitors with varying silicon content (0 cat% – 8.5 cat%). All samples were annealed at 1000 °C for 1 s. Transitions from paraelectric to ferroelectric and from ferroelectric to antiferroelectric-like behaviour are visible for increasing silicon content.

commonly include charged oxygen vacancies. In the presence of these defect dipoles, the existing multi-domain structure becomes stabilised, whereas the domain wall movements are impeded [56], [208], [209]. In this case the crystalline structure of the material remains the same for all film compositions. It is the domain switching capability, which undergoes altering. On the contrary, in the ferroelectrics with the 1st order phase transition double hysteresis loops observed slightly above their Curie point were attributed to a field-induced transition of a paraelectric into a ferroelectric phase [210], [77]. Therefore, antiferroelectric-like features can be observed close to a phase transition, where equilibrium between two phases can be affected by electric field. In order to get better insight into the origin of the ferroelectric and antiferroelectric-like behaviour of Si:HfO$_2$ system, microstructure of films with different silicon content was examined by means of grazing incidence XRD and high-resolution TEM. The results of the microstructural analyses and their correlation to electrical film properties will be discussed in the next section 5.1.2.

5.1.2   Structural characterisation

The high-resolution TEM analysis (chapter 3.4.3) was used to obtain information about the microstructure of crystalline Si:HfO$_2$ films. Samples containing 27 nm thick Si:HfO$_2$ layers with 4.4 cat% Si content annealed at 800 °C were chosen as a representative of structures with ferroelectric behaviour. Figure 5.2 shows the resulting TEM images from different sample areas. It can be seen, that the studied Si:HfO$_2$ films exhibited a polycrystalline nature with an average grain size of 20 – 30 nm and predominantly columnar grain morphology. The height of the grains was approximately equal to the film thickness. The films under the Pt dots embedded between TiN layers (Figure 5.2 (b)) and near the Pt dots, with wet-etched TiN top electrodes – bare Si:HfO$_2$ surface (Figure 5.2 (c)) showed similar microstructure.

The impact of silicon content on the phase composition of Si:HfO$_2$ films was investigated by GI-XRD (chapter 3.4.1). Figure 5.3 (b) shows the GI-XRD scans for samples with varying Si doping levels after anneal at 650 ºC for 20 s. The films with the silicon content above 5.6 cat% remained amorphous, while pure HfO$_2$ films were almost completely crystalline. Therefore, it can be deduced, that the incorporation of silicon resulted in an increase of the HfO$_2$ crystallisation temperature. A similar effect of increasing crystallisation temperature by doping with Si, Al and La was reported for HfO$_2$ ceramics [120] as well as thin films [129], [211] – [214].

Furthermore, Si-doping induced a change in the crystalline structure, which became more evident for samples annealed at higher temperatures. Figure 5.4 shows the GI-XRD diffractograms for 9 nm Si:HfO$_2$ films with Si content of 0, 4.4 and 8.5 cat% after anneal

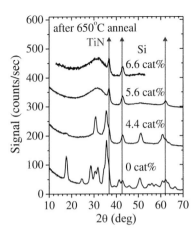

Figure 5.2 (a) High-resolution TEM image showing Pt/TiN/Si:HfO₂/TiN/Si capacitor stack. Enlarged images of the stack (b) under a Pt dot and (c) next to Pt dot, with wet-etched TiN top layer. Samples under test contain 27 nm Si:HfO₂ with 4.4 cat% Si annealed at 800 °C for 20 s.

Figure 5.3 Experimental GI-XRD diffractograms for 9 nm Si:HfO₂ films with varying silicon content after anneal at 650 °C for 20 s. An increase in the silicon content induced a rise of the crystallisation temperature and change in the crystalline structure. Reference powder diffraction pattern of the cubic TiN (Fm3m) is taken from [203].

at 1000 °C for 1 s. The reference powder diffraction patterns for tetragonal (P4₂/nmc) [202], orthorhombic[1] (Pbc2₁) [215] and monoclinic (P2₁/c) [216] phases are also introduced for comparison. The difference in the peak positions between experimental scans and reference patterns in corresponding phases can arise from internal lattice strains in the studied films or/and slightly different lattice parameters in respect to the reference powder samples. A visible change in the XRD patterns occurred with increasing Si concentration. Pure HfO₂ films crystallised predominantly into the monoclinic phase with characteristic (-111) and (111) peaks at $2\theta = 28.5°$ and $31.8°$ and a small fraction of the tetragonal crystallites[2] (peak at $2\theta = 30.8°$). The stability of the tetragonal phase increased for samples with higher Si concentration. The (101) peak at $2\theta = 30.8°$ stood out more, while the monoclinic peaks decreased in intensity until they completely vanished for Si content above 5.6 cat%.

---

[1] No reference pattern of the orthorhombic (Pbc2₁) phase for HfO₂ was available. The XRD data bases include patterns, acquired from bulk materials and powders, where this phase was never observed. It was only recently discovered in thin films [18]. Therefore, a diffraction patterns for the orthorhombic (Pbc2₁) phase in ZrO₂, which exhibits a crystalline structure almost identical to HfO₂ [117], was used here as a reference.

[2] This peak should not appear in the monoclinic symmetry and evidences a presence of a different phase. The exact identification of this phase, however, is difficult due to its small volume fraction.

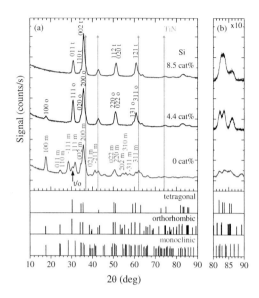

Figure 5.4 (a) GI-XRD diffraction scans for 9 nm Si:HfO$_2$ films after annealing at 1000 °C for 1 s, containing 0 cat% (pure HfO$_2$), 4.4 cat% and 8.5 cat % silicon. Reference powder diffraction patterns for the tetragonal (P4$_2$/nmc) [202], orthorhombic (Pbc2$_1$) [215], and monoclinic (P2$_1$/c) [216] HfO$_2$ phases as well as the cubic TiN (Fm3m) [203] are shown for comparison. (b) Enlarged GI-XRD patterns (factor 10) for the 2θ range between 80° and 90°.

In bulk HfO$_2$ materials the tetragonal phase is known to be stable only for temperatures above 1700 °C [117]. In thin HfO$_2$ films, on the contrary, this high-temperature tetragonal phase can be observed at significantly lower temperatures. This effect was attributed to the increased impact of surface energy term, which becomes comparable to the volume energy term, and starts to impact the material properties [120]. Doping of HfO$_2$ with Si [211], [123], [217] and several other elements [125], [126], [218] was reported to facilitate crystallisation into the tetragonal phase, which is consistent with our results. At Si content of 4.4 cat% the phase transition between monoclinic and tetragonal phases took place. The peak at 30.8° associated with the tetragonal phase was observed together with the peak at 17.6°, indicating the presence of a low-symmetry phase. The orthorhombic phase (Pbc2$_1$) is expected to appear at this phase boundary between the tetragonal and the monoclinic phases [18], [135] as a result of a martensitic phase transformation from the metastable tetragonal phase. As argued by Kisi in [136] a tetragonal-to-monoclinic phase transition becomes unfavourable in the presence of internal lattice strains, because of the required volume expansion of about 3.5% [117] in combination with shearing and twinning of a unit cell. As a result, a tetragonal-to-orthorhombic transition with less volume expansion and shear less unit cell transformation was argued to proceed instead. The internal film stains in our case were induced by the TiN top and bottom electrodes, present during the film crystallisation. The appearance of the orthorhombic phase ($Pbc2_1$) for HfO_2 films containing 4.4 cat% Si provides a good explanation of the ferroelectric properties detected for this film composition (Figure 5.1). This phase is the only non-centrosymmetric crystalline phase known for HfO_2 and, therefore, the only phase, which meets the requirements of ferroelectricity. The identification of the orthorhombic phase is, however, a difficult task, since its XRD-patterns is very similar to that of a monoclinic/tetragonal phase mixture. The major XRD reflexes from these phases appear at similar diffraction angles (Table 5.1). In addition, the textured structure[3] of the studied samples made the phase assignment even more difficult. Nevertheless, by using Rietveld refinement [199], the high-pressure orthorhombic (Pbcm) phase highly textured in (100) direction in combination with (001) textured tetragonal ($P4_2$/nmc) phase was identified for samples with 4.4 cat% Si. The Pbcm orthorhombic phase is centrosymmetric and cannot explain the observed ferroelectric behaviour. However, unit cell structures in Pbcm and expected $Pbc2_1$ phases are related and cannot be distinguished by means of XRD [220].

Table 5.1 Diffraction peak positions determined from experimental GI-XRD patterns for HfO_2 films with 4.4 cat% silicon and the theoretical diffraction reflexes for the monoclinic ($P2_1$/c), tetragonal ($P4_2$/nmc) and orthorhombic ($Pbc2_1$) HfO_2 phases that can be assigned to the experimental peak positions.

Peak position (2θ)	Monoclinic ($P2_1$/c)	Tetragonal ($P4_2$/nmc)	Orthorhombic ($Pbc2_1$)
17.6°	(100)	–	(100)
24.6°	(011)/(110)	–	(110)
28.5°	(-111)	–	–
30.8°	–	(101)	(111)
31.8°	(111)	–	–
35.6°	(200)	(110)	(200)

---

[3] Texture with {100} orientation was also observed up to different extent for HfO_2 films grown from metal-organic precursors [18], [130]. Therefore, the observed texture is not a specific feature of the chlorine-based ALD process, but was rather determined by the substrate structure [219], in our case – by the structure of the bottom TiN layer.

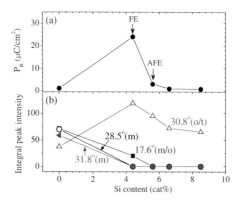

Figure 5.5 Correlation between the Si content dependence of (a) the remanent polarisation (P_R) and (b) the integral intensities of dominant XRD peaks extracted from GI-XRD for 9 nm Si:HfO$_2$ annealed at 1000 °C for 1 s (Figure 5.4 (a)). The highest P_R values was detected for films with 4.4 cat% Si, which corresponded to the monoclinic-to-tetragonal phase.

Therefore, the assumption of the presence of the non-centrosymmetric Pbc2$_1$ phase for this film composition is fully justified. An additional proof of the presence of a new crystalline phase for Si:HfO$_2$ with 4.4 cat% Si besides monoclinic and tetragonal provided the GI-XRD scans measured in the 2θ range between 80° and 90° (Figure 5.4 (b)). The XRD pattern for 4.4 cat% Si cannot be reproduced by a superposition of the upper and lower graphs, which would correspond to a simple phase mixture of the monoclinic and the tetragonal phases. From the comparison of the GI-XRD scans (Figure 5.4) with the results of the electrical measurements (Figure 5.1) a clear correlation between the crystalline structure of Si:HfO$_2$ films and their electrical properties can be derived. The observed change in the electrical behaviour was a result of the phase transition induced by silicon doping. The ferroelectric behaviour was detected at the phase transition between the monoclinic (P2$_1$/c) and the tetragonal (P4$_2$/nmc) phases, where the non-centrosymmetric orthorhombic Pbc2$_1$ phase emerged. An appearance of antiferroelectric-like characteristics was directly related to the stabilisation of the tetragonal phase. Figure 5.5 shows the Si-content dependencies of the remanent polarisation (P_R), obtained from the P-V measurements (Figure 5.1 (a)), and the integral intensity of several major XRD reflexes extracted from the GI-XRD scans (Figure 5.4 (a)). The P_R maximum was detected for films with 4.4 cat% Si, which corresponded to the monoclinic-to-tetragonal phase transition. An increase in the silicon content above 4.4 cat% Si was accompanied by a significant decrease of P_R. Another point worth mentioning is the alteration of the integral intensity of the diffraction peak at $2\theta = 30.8°$, which showed similar dependence to P_R and dropped for Si content above

4.4 cat%. This peak is characteristic for both the tetragonal (P4$_2$/nmc) and the orthorhombic (Pbc2$_1$) phases. The decrease of its intensity, therefore, indicated the reduction in the fraction of one of the phases – in our case the orthorhombic, since the tetragonal should become more stable. The residuals of the orthorhombic phase could be the cause of the observed double hysteresis loops for films with Si content > 4.4 cat% (Figure 5.1 (a)). A field-induced phase transition between from the nonpolar tetragonal (P4$_2$/nmc) and polar orthorhombic (Pbc2$_1$) phases can be another possible explanation of the antiferroelectric characteristics. The possibility of such phase transitions was recently shown on the basis of first principle calculations for HfO$_2$ [221] as well as ZrO$_2$ [222], which is chemically and structurally similar to HfO$_2$.

5.1.3 Piezoresponse force microscopy analysis

Piezoelectricity is a mandatory property of ferroelectric materials (see discussion to Figure 2.8). Therefore, in order to confirm a true ferroelectric nature of polarisation hysteresis observed for Si:HfO$_2$ films, their piezoelectric properties were studied by piezoresponse force microscopy (PFM) (chapter 3.2.4). All measurements were performed on crystalline 9 nm thick Si:HfO$_2$ films (after a 1000 °C anneal), exhibiting distinct polarisation hysteresis loops.

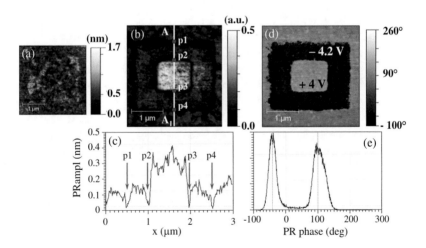

Figure 5.6 PFM results for 9 nm thick ferroelectric Si:HfO$_2$: (a) Topography map; (b) piezoresponse amplitude map; (c) *PRampl* profile along AA$_1$ section; (d) piezoresponse phase map; (d) histogram of the *PRphase*.

The local piezoelectric response from domains with opposite polarisation on a free Si:HfO$_2$ surface was investigated at first. Domain square polarisation patterns were written by applying constant poling voltages of -4.2 V and $+4$ V to the probe tip, while scanning the film surface with decreasing area (2.5×2.5 μm^2 and 1×1 . μm^2, respectively). The piezoresponse images were subsequently acquired using AC voltage with amplitude of 1 V. The resulting maps of the out-of-plane piezoresponse amplitude (*PRampl*) and phase (*PRphase*) together with a topography map are depicted in Figure 5.6. The polarised square areas were clearly observed in the *PRampl* (Figure 5.6 (b)) and the *PRphase* (Figure 5.6 (d)) images, whereas the surface topography did not exhibit a comparable pattern (Figure 5.6 (a)). The boundaries to the outer unpolarised area as well as between the oppositely polarised square areas corresponded to regions with zero *PRampl* (black lines), which is an indicator of domain boundaries. The *PRampl* profile (Figure 5.6 (c)) along the section *AA$_1$* showed four minima (*p1-p4*), i.e. four domain boundaries. The *PRphase* jumps were found at the same positions (Figure 5.6 (d)). The dark and bright areas with an average *PRphase* difference of $140°$ represented regions with opposite orientations of polarisation (up and down polarised domains, respectively). The presence of piezoresponse and the demonstrated ability to reverse the film polarisation by applying an external electric field serve as a strong evidence of intrinsic ferroelectric behaviour in Si:HfO$_2$ films [223]. From the sharp domain boundaries observed with PFM and a predominantly columnar film structure detected by TEM (Figure 5.2) it can be concluded, that the ferroelectric domains extended through the entire film as stated in [173]. Moreover, the areas polarised during PFM analysis (several micrometres) must have consisted of multiple grains, since they were significantly larger than an average grain size ($20 - 30$ nm) estimated by TEM (Figure 5.2).

Furthermore, the local switching behaviour was studied by measuring piezoelectric response as a function of bias voltage at different sample locations in both operation modes: in-field, with DC bias simultaneously applied, (Figure 5.7) and off-field, at 0 V after a specific bias was turned off, (Figure 5.8). These measurements were performed on Si:HfO$_2$ capacitors with top electrodes, which were previously bonded. This enabled to decouple the DC bias (applied to the top electrode) from the AC sensing signal at the probe tip in order to minimise electrostatic effects and generate a homogeneous electric field in the films [223]. The hysteresis data were collected within the 3×3 μm^2 area. Figure 5.7 and Figure 5.8 show in-field and off-field characteristics for a specific position on the sample. Similar curves with slight variations in the values of the piezoresponse were obtained for other positions. For both measurement modes the *PRphase* as well as piezoresponse characteristics showed a hysteretic behaviour, indicating a polarisation switching during bias sweeps.

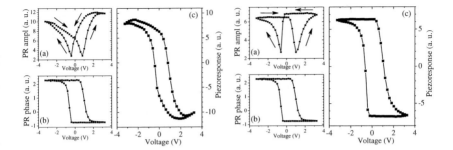

Figure 5.7 In-field piezoresponse-voltage measurement: (a) *PR* amplitude, (b) *PR* phase and (c) piezoresponse.

Figure 5.8 Off-field piezoresponse-voltage measurement: (a) *PR* amplitude, (b) *PR* phase and (c) piezoresponse.

The off-field measurements indicated the retaining of the induced polarisation states even in the absence of external fields. Therefore, they served as an evidence of good local retention properties of the ferroelectric Si:HfO$_2$. The butterfly-shaped curves were acquired for the *PRampl*, which is a typical feature of ferroelectric materials [56], [77]. The electrostrictive effect during the in-field measurements resulted in additional voltage dependence of the *PRampl*, not present in the off-field characteristics. Here, constant *PRampl* values during the back sweeps (from +/ – 3 V to 0 V) emphasised again good polarisation retention properties. The minima of these characteristics corresponded to the polarisation reversal and, thus, provided the values of the local coercive fields. An average value of the local E_C, extracted from PFM test, was about 1 MV/cm, which is in good agreement with macroscopic value obtained from *P-V* measurements (chapter 5.1.1).

5.2   Impact of the post-metallisation anneal

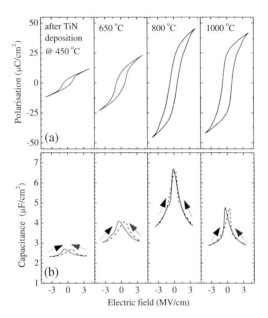

Figure 5.9 Impact of annealing temperature on (a) the polarization- and (b) capacitance-voltage characteristics of ferroelectric capacitors containing 9 nm Si:HfO₂ films with silicon contents of 4.4 cat%. Ferroelectric behaviour remained for all annealing conditions.

The impact of post-metallisation anneal on the ferroelectric properties of Si:HfO₂ layers was investigated. Electrical properties and crystalline structure were analysed for 9 nm films exposed to different post-metallisation annealing conditions: 650 °C for 20 s, 800 °C for 20 s and 1000 °C for 1 s as well as the as-deposited case without additional thermal treatment besides the thermal budget of 450 °C for 6 – 8 hours during the deposition of TiN top electrodes.

The electrical characterisation was performed using the polarisation- and capacitance-voltage measurements. Figure 5.9 shows the resulting characteristics for the films with 4.4 cat% silicon, which were previously found to exhibit ferroelectric behaviour (section 5.1.1). Polarisation hysteresis loops in combination with butterfly-shaped *C-V* curves were detected for all annealing conditions including the as-deposited case. The similarity of electric properties for annealed and as-deposited sample indicated partial crystallisation of Si:HfO₂ films during the deposition of TiN top electrode, where the films were exposed to

450 °C for 6 – 8 hours. This assumption was also confirmed by the results of the GI-XRD measurements (Figure 5.10 (a)). Here, diffraction peaks related to crystalline HfO_2 at $2\theta = 30.8°$ and 35.6 ° were detected. Exposure of samples to additional post-metallisation anneal resulted in more apparent ferroelectric behaviour with larger hysteresis loops and higher P_R-values (Figure 5.9). The number and positions of the diffraction peaks remained unchanged with increasing annealing temperature (Figure 5.10 (a)). All observed peaks could be assigned to the orthorhombic (Pbc2$_1$) phase. The primary effect of the higher annealing temperatures consisted in an increase in the integral intensities of all HfO_2 peaks (Figure 5.10 (b)), which indicated a growth in the degree of film crystallinity. Moreover, a slight decrease in the FWHM (full width at half maximum) values was detected for higher annealing temperatures, which evidenced an increase in the grain size. The intensity of the TiN peak at $2\theta = 42.8°$, on the contrary, remained constant for all annealing conditions, since the TiN was already crystalline after deposition and its crystallinity did not change with further annealing. Thus, the TiN peaks were used as a reference to evidence the change in the crystallinity of HfO_2 films. The remanent polarisation showed a monotonic increase with the annealing temperature (Figure 5.10 (c)), similar to the integral intensity of the HfO_2 diffraction peaks (Figure 5.10 (b)). The coercive field strength, on the other hand, was almost independent on the annealing conditions. Therefore, an altered degree of film crystallinity, which is directly related to the fraction of the ferroelectric phase, can be held responsible for the enhancement of the ferroelectric properties at higher annealing temperatures. Special attention should be given to the fact, that annealing at 1000 °C for 1 s was enough to crystallise the Si:HfO_2 films and to induce their ferroelectric properties with $P_R = 24$ µC/cm^2.

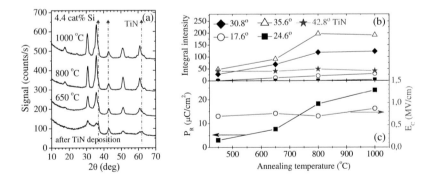

Figure 5.10 (a) GI-XRD scans of 9 nm thick ferroelectric Si:HfO_2 films (4.4 cat% Si) after different annealing treatments. Reference powder XRD patterns for the cubic TiN (Fm3m) were taken from [203]; Integral intensities of major XRD peaks (b), the remanent polarisation P_R and the coercive field E_C (c) versus annealing temperature.

This finding is of significance for later integration of Si:HfO$_2$ films into FeFET memory devices. A spike anneal with similar parameters is used for CMOS transistors for dopant activation. For Si:HfO$_2$-based FeFETs it can be simultaneously used to crystallise Si:HfO$_2$ films into a ferroelectric phase. Therefore, Si:HfO$_2$-based ferroelectric transistors can be fabricated with the state-of-the-art CMOS process without any additional annealing steps needed.

As was shown in chapter 5.1, ferroelectric properties of Si:HfO$_2$ emerged on the tetragonal-to-monoclinic phase boundary, which is induced by Si doping. Impact of the thermal treatment on the film composition exhibiting ferroelectric behaviour was examined. For this purpose P-V characteristics were recorded for samples with varying silicon content and different annealing conditions. The extracted P_R-values are shown as a function of silicon content in Figure 5.11. Independent of annealing conditions ferroelectric behaviour appeared for the film composition with 4.4 cat% Si, where the maximum P_R was detected (Figure 5.11). At all annealing temperatures this film composition also corresponded to the monoclinic-to-tetragonal phase transition similar to Figure 5.3. Therefore, it can be deduced, that the film crystalline phase formed during crystallisation was predominantly determined by the film composition, whereas the annealing conditions affected a degree of the film crystallinity. The emergence of ferroelectric properties will, thus, depend on the ability to control the film composition. In Si:HfO$_2$ films the ferroelectric properties were detected in a rather narrow range of low silicon content (2.6 – 4.9 cat% [18]), where a precise tuning of film composition and, hence, film properties is rather difficult. Of great interest are, therefore, dopants, which provide ferroelectric properties in HfO$_2$ within a broader doping range (e.g. Zr (30-70 cat%) [132] and Sr [224]).

Leakage current behaviour is one of the reliability concerns in the thin films, which is also relevant for ferroelectric memory applications. High leakage currents cause a malfunction in both types of ferroelectric memory cells, degrading reliability for FeRAM [49] and retention for FeFET [110] devices. Influence of annealing conditions on the leakage current characteristics was examined for 9 nm thick ferroelectric Si:HfO$_2$ films with 4.4 cat% Si (Figure 5.12). An increase in the leakage current with annealing temperature was detected for both voltage polarities. Crystallisation was previously identified as the main source of the leakage current degradation in HfO$_2$ films [214], [217], which also coincides well with observations made for studied Si:HfO$_2$ films. Grain boundaries were shown to serve as high leakage paths in crystalline films [225].

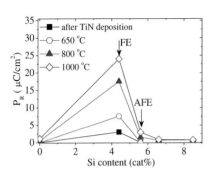 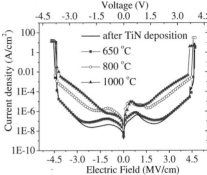

Figure 5.11 P_R versus Si content dependence for 9 nm thick Si:HfO$_2$ films exposed to different annealing conditions [23].

Figure 5.12 Impact of annealing temperature on the leakage current characteristics of 9 nm thick ferroelectric Si:HfO$_2$ films (4.4 cat% Si).

5.3 Impact of the film thickness

Properties of thin dielectric films can be essentially affected by their thickness [217], [127]. In thin films the contribution from the surface energy becomes comparable to the volume energy due to a high surface-to-volume ratio of each individual grain [120] and starts to affect their physical properties. The grain size and, thus, the surface-to-volume ratio correlate directly with the film thickness. Therefore, the film properties can be shifted in the direction of either bulk-dominated or surface-dominated by adjusting their thickness. In this chapter the influence of thickness on the electrical properties and crystallisation behaviour of Si:HfO$_2$ films will be discussed. MIM capacitors including 9, 27 and 50 nm thick Si:HfO$_2$ layers were studied. It will be shown, that the crystallisation behaviour of the Si:HfO$_2$ films was altered with increase in the film thickness, which in turn influenced the ferroelectric characteristics.

The impact of silicon content and annealing temperature on the crystalline structure of Si:HfO$_2$ films with varying thickness (9, 27 and 50 nm) was investigated by means of GI-XRD. The crystal structure of the layers was identified by comparing the experimental scans with theoretical powder XRD patterns of the monoclinic (P2$_1$/c) [216], tetragonal (P4$_2$/nmc) [202] and orthorhombic (Pbc2$_1$) [215] phases of HfO$_2$. Figure 5.13 (a) and (b) summarizes the results of the structural analysis for 9 and 27 nm thick Si:HfO$_2$ films, respectively. (The phase diagrams of 27 and 50 nm films were identical.) For all film thicknesses the incorporation of silicon had the same effect and resulted in increase

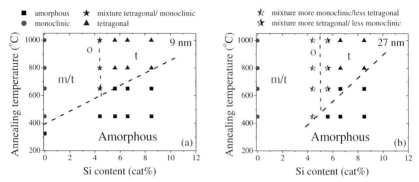

Figure 5.13 Effect of silicon content and annealing temperature on the phase composition of (a) 9 nm and (b) 27 nm thick Si:HfO$_2$ films (similar to 50 nm films) [23]. For both film thicknesses an increase in the silicon content induced rise in the crystallisation temperature in combination with stabilisation of the tetragonal phase. With increasing film thickness the monoclinic-to-tetragonal transition shifted to slightly higher Si content.

of the crystallisation temperature and stabilisation of the tetragonal phase. With growing film thickness, however, the crystallisation temperature decreased in comparison to 9 nm films with the same silicon content (e.g. 4.4 and 5.6 cat% silicon). The GI-XRD diffractograms of the films with 4.4 cat% Si and varying film thickness measured directly after the deposition of the TiN top electrodes at 450$^°$C are shown in Figure 5.14 (a). The 9 nm films remained predominantly amorphous. On the contrary, for 27 and 50 nm films a crystalline structure was detected, indicating at least partial crystallisation. Similar impact of the film thickness on the crystallisation temperature was previously observed for pure HfO$_2$ films [226], [227] as well as for HfSiON films [127]. Increase in the film thickness, moreover, enhanced the stability of the monoclinic phase. The characteristic monoclinic peaks (-111) and (111) at $2\theta = 28.5$$^°$ and 31.8$^°$, respectively, were clearly observed for the 27 nm and 50 nm films, whereas no evidence of these peaks was found for 9 nm films (Figure 5.14 (a)). An increased film thickness resulted in a shift of the monoclinic-to-tetragonal phase transition to slightly higher silicon content: from 4.4 cat % Si for 9 nm films to some concentration between 4.4 and 5.6 cat% Si for thicker films. In the GI-XRD scans obtained for 27 nm Si:HfO$_2$ film with varying composition (Figure 5.14 (b)), the residuals of the monoclinic phase was detected up to 5.6 cat% Si. Both effects – decrease of the crystallisation temperature and increased stability of the monoclinic phase with increasing film thickness – can be explained by the surface energy effect [120], [121]. In the thin films with high surface-to-volume ratio the properties are affected by the volume as well as surface energy. The impact of the surface energy diminishes as the films become thicker. As a result the properties of thick films approach those of the bulk material. Therefore, in the case of thick films higher silicon doping was needed to induce the monoclinic-to-tetragonal phase transition as well as to prevent film crystallisation.

The electrical properties of Si:HfO$_2$ films were expected to vary with the film thickness, since it was found to affect their crystalline structure. *P-V* and *C-V* measurements were performed on films with different thickness. The results of electrical characterisation for 9 nm films were discussed in sections 5.1 and 5.2. The ferroelectric behaviour was shown to appear for Si content of 4.4 cat%, corresponding to the monoclinic-to-tetragonal phase boundary. For thicker films the phase transition was shifted to slightly higher Si content (see Figure 5.13 (b) and Figure 5.14 (b)): between 4.4 and 5.6 cat% Si. Hysteretic *P-V* curves were obtained for both these compositions (Figure 5.15 after TiN deposition and 650 °C anneal), confirming again, that the ferroelectric properties in Si:HfO$_2$ films emerge at the phase boundary. The temperature dependence of the *P-V* characteristics varied, however, with the silicon doping level and showed a completely different trend in comparison to the 9 nm films (chapter 5.2). *P-V* characteristics of 27 nm films with 4.4 cat% Si remained unaffected by the thermal

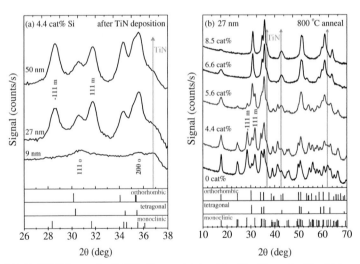

Figure 5.14 (a) GI-XRD scans for Si:HfO$_2$ films containing 4.4 cat% with varying film thickness after deposition of the TiN top electrodes; (b) GI-XRD scans for 27 nm Si:HfO$_2$ films with different composition after annealing at 800°C for 20 s [23]. Growth of the film thickness resulted in an increased stability of the monoclinic phase. Reference powder XRD patterns: the orthorhombic (Pbc2$_1$) [215], tetragonal (P4$_2$/nmc) [202] and monoclinic (P2$_1$/c) [216] phases of HfO$_2$ and the cubic TiN (Fm3m) [203].

treatment. For the 5.6 cat% Si slight hysteresis pinching appeared for 650 °C anneal and transformed into antiferroelectric-like double hysteresis loops with further increase in the annealing temperature.

The altered impact of thermal treatment on the ferroelectric characteristics of the thick Si:HfO₂ films was studied in more details. In order to get better insight into its origin, the correlation between the *P-V* measurements and crystalline structure was analysed for 27 nm thick Si:HfO₂ films with 4.4 cat% Si. For these samples the post-metallisation treatment had a negligible effect of on the polarisation hysteresis loops (Figure 5.15(a)). Comparable P_R-values were extracted for all annealing conditions (Figure 5.16 (a)). A contrary observation was made for 9 nm films, where an apparent increase in the P_R with the annealing temperature was detected (Figure 5.10). This was found to be related to the growing degree of the film crystallinity. The analyses of the film crystallinity for 27 nm films were performed using GI-XRD measurements. Figure 5.16 (b) depicts the integral intensities of dominant XRD peaks as a function of annealing temperature. Again a good correlation between the annealing temperature dependence of P_R and the integral intensities of HfO₂ peaks could be observed. Only minor change in the peak intensities for all HfO₂ peaks occurred for varying annealing temperatures. This indicates, that 27 nm films crystallised already during the deposition of the TiN top electrode at 450 °C, which is consistent with the detected lowering of their crystallisation temperature. The subsequent annealing step changed neither the degree of crystallinity nor the phase composition. Therefore, the P_R-value remained independent from the annealing temperature.

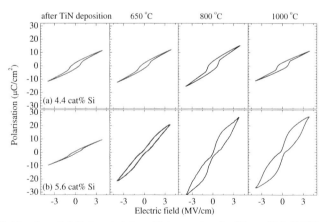

Figure 5.15 Alteration of *P-V* characteristics with temperature for 27 nm thick Si:HfO₂ films with compositions at the monoclinic-to-tetragonal phase boundary (a) 4.4 cat% and (b) 5.6 cat% silicon.

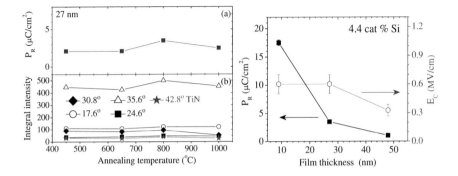

Figure 5.16 Correlation between the temperature dependence of (a) the remanent polarization and (b) integral intensities of XRD for 27 nm thick Si:HfO₂ films with 4.4 cat% Si.

Figure 5.17 The impact of film thickness on the remanent polarisation (P_R) and the coercive field strength (E_C) of Si:HfO₂ with 4.4 cat% Si annealed at 800 °C for 20 s.

Furthermore, an increase in the film thickness resulted in a significantly decreased remanent polarisation (Figure 5.16): from 18 μC/cm^2 for 9 nm films down to 1 μC/cm^2 for 50 nm films. The E_C-values, on the contrary, were comparable for 9 and 27 nm layers, decreased, however, for 50 nm layers. This suppressed ferroelectric behaviour in thick films was probably a result of the increased stability of the monoclinic phase (Figure 5.14 (b)), which inhibited the formation of the ferroelectric orthorhombic phase. Two possible reasons could cause the observed stabilisation of the monoclinic phase with increasing thickness: decreased impact of the surface energy contribution, causing a transition to the volume dominated properties [120], and also insufficient mechanical stress during crystallisation. The 27 and 50 nm films crystallised already during the deposition of the TiN electrodes. Films with 9 nm thickness, on the contrary, remained predominantly amorphous during metallisation and were crystallised in a subsequent annealing step in the presence of both TiN electrodes, thus, under mechanical confinement. The role of mechanical confinement is not completely understood yet. It was reported to be of secondary importance for the formation of the ferroelectric phase in Al-doped HfO₂ system [130]. For the pure HfO₂ [128] and Si:HfO₂ [18], however, the presence of a TiN capping layer was found to facilitate the stabilisation of the tetragonal/orthorhombic phase. Therefore, ferroelectric properties in 27 – 50 nm thick Si:HfO₂ films can be enhanced, if they crystallise in the presence of mechanical confinement. This assumption was recently confirmed in [228].

5.4   Summary

The material aspects of Si-doped HfO$_2$ thin films have been studied in order to gain better insight into the occurrence of ferroelectricity in this system and to acquire guidelines for transistor fabrication. The influence of the different process parameters such as the Si doping concentration (chapter 5.1), post-metallisation annealing conditions (chapter 5.2) and film thickness (chapter 5.3) on the stabilisation of the ferroelectric properties in Si:HfO$_2$ films has been examined. Electrical characterisation combined with structural analyses enabled the changes in the macroscopic electrical properties to be correlated to alterations in the film crystalline structure.

The film composition appears to have a strong impact on the electrical properties of the Si:HfO$_2$ films (Figure 5.1). By varying the Si doping level the electrical properties could be tuned between paraelectric (pure HfO$_2$), ferroelectric (4.4 cat% Si) and antiferroelectric-like ($\geq$ 5.6 cat% Si) behaviour. In this case, the Si doping has two effects; it increases the film crystallisation temperature (Figure 5.3) and simultaneously induces a phase transition from the monoclinic (P2$_1$/c) to the tetragonal (P4$_2$/nmc) phase (Figure 5.4). A combination of both these effects resulted in the observed ferroelectric as well as antiferroelectric-like behaviour. As a result of the increase of crystallisation temperature with doping, Si:HfO$_2$ could be encapsulated in an amorphous phase and crystallised in the presence of both TiN electrode layers that serve as a mechanical confinement. This confinement facilitated the formation of a ferroelectric phase at the monoclinic-to-tetragonal phase boundary (Figure 5.4 and Figure 5.5). The effect of the Si-doping was similar to other dopants, for which ferroelectric behaviour has also been observed at the monoclinic-to-tetragonal phase boundary [19], [20], [130], [131]. It has been argued that crystallisation in the presence of mechanical confinement is a prerequisite for the formation of the ferroelectric phase in the Si:HfO$_2$ material system [18]. This statement could be partially confirmed in this work (chapter 5.3). For other doping elements (e.g. Al [130]) this condition was, however, shown to be less important. A non-centrosymmetric orthorhombic phase Pbc2$_1$, which can be stabilised at the monoclinic-to-tetragonal phase boundary as claimed in [18], was also held responsible for the ferroelectric properties of films studied in this work. An unambiguous identification of the orthorhombic Pbc2$_1$ phase was unfortunately impossible. The presence of a new crystalline HfO$_2$ phase with a symmetry different from the monoclinic (P2$_1$/c) and tetragonal (P4$_2$/nmc) phases was, however, ascertained at the monoclinic-to-tetragonal phase boundary by means of GI-XRD analyses. In addition to the *P-V* hysteresis loops and butterfly-shaped *C-V* curves, a strong evidence for the structural ferroelectricity in Si:HfO$_2$ films (i.e. existence of a non-centrosymmetric ferroelectric phase) was provided by PFM measurements (chapter 5.1.3). A distinct piezoelectric response, which is a necessary requirement of structural ferroelectricity

[69], was demonstrated for the Si:HfO$_2$ films in combination with the ability to locally reverse the film polarisation in an external electric field. Nevertheless, other origins of the observed ferroelectric behaviour in Si:HfO$_2$ films (e.g. oxygen vacancy driven ferroelectricity) cannot be completely ruled out. In order to clarify the nature of the ferroelectricity in HfO$_2$-based materials, ab initio structural simulations and further experimental studies are required.

The electrical characteristics of ferroelectric Si:HfO$_2$ thin films is of greater relevance than a complete understanding of the root cause of their ferroelectricity for application in ferroelectric memories. The studied Si:HfO$_2$ layers exhibited P_R-values (18 – 24 μC/cm^2) comparable to those of perovskite-type ferroelectric materials (e.g. PZT and SBT (see Table 2.1)) and about a factor of ten higher E_C of ~ 1 MV/cm. The latter is advantageous for FeFET cell-types, where the gate stack height can be reduced to a few nanometres and still exhibit a sufficiently high memory window, in contrast to devices with PZT and SBT films [140]. Therefore, Si:HfO$_2$ ferroelectrics provide better scaling potential for FeFET cells. Utilisation of chloride-based precursors and H$_2$O for the fabrication of Si:HfO$_2$ films allowed higher P_R-values to be achieved in comparison to films grown with metal-organic precursors and O$_3$ that exhibited P_R-values from 5 to 12 μC/cm^2 [22], [207], [229]. The values of the coercive fields (0.7 – 1 MV/cm) and the Si doping range with the most prominent ferroelectric properties (2.5 – 4 cat% Si) were, on the other hand, comparable for both deposition processes. Chloride-based precursors are typically associated with an increased level of chlorine contamination in the grown films as well as a reduced carbon contamination level [230], [231], which may have affected the electrical properties of the films.

The appearance of antiferroelectric-like characteristics in Si:HfO$_2$ films was directly related to the stabilisation of the tetragonal phase with increasing Si content (Figure 5.5). The origin of the antiferroelectric behaviour is yet not completely understood. The most plausible explanations are the residuals of the orthorhombic phase presented in films with Si content greater than 4.4 cat% as well as a field-induced phase transition between the tetragonal (P4$_2$/nmc) and the orthorhombic (Pbc2$_1$) phases. The possibility for the latter explanation was recently confirmed on the basis of first principle calculations for both HfO$_2$ [221] and ZrO$_2$ [222].

The post-metallisation anneal affected primarily the degree of the film crystallinity and had no impact on the position of the monoclinic-to-tetragonal phase boundary. Annealing at higher temperatures resulted in increased P_R-values (Figure 5.10), due to an increased degree of film crystallinity, which was directly related to the fraction of the ferroelectric phase. A side-effect of the increased film crystallinity with annealing temperature was a higher leakage current (Figure 5.12). Annealing at 1000 °C for 1 s, equivalent to dopant activation anneal used during CMOS process, was shown to be sufficient for crystallisation of Si:HfO$_2$ films and to obtain a P_R of 24 µC/cm^2. Therefore, Si:HfO$_2$-based ferroelectric transistors can be fabricated using state-of-the-art CMOS process without a requirement for additional annealing steps. This is a real advantage in comparison to the PZT and SBT films that require special integration schemes due to the high processing temperatures (600 – 800 °C), high pressure oxygen atmosphere during deposition and the sensitivity of the ferroelectric properties to the hydrogen used during forming gas anneals [97], [98], [17].

The ferroelectric behaviour of Si:HfO$_2$ also depends on the film thickness. The increased stability of the monoclinic phase for thicker films (Figure 5.14) impeded the formation of a ferroelectric orthorhombic phase, leading to reduction in the remanent polarisation (Figure 5.17). The observed stabilisation of the monoclinic phase for thicker films resulted from the combined effect of the decreased surface energy [120], [121] and insufficient mechanical stress during crystallisation. Thicker films (27 and 50 nm) crystallised already during the deposition of the top TiN electrode due to a reduced crystallisation temperature, while the thinner films (9 nm) were crystallised after they were embedded between top and bottom TiN layers.

6 Electrical properties of the ferroelectric Si:HfO₂ thin films

The application of Si:HfO$_2$ films in ferroelectric memories requires a good knowledge of their electrical behaviour. Therefore, this chapter focuses on the electrical properties of ferroelectric Si:HfO$_2$ films, which are relevant for memory applications: effect of the field cycling (section 6.1), polarisation switching speed (section 6.2) and fatigue characteristics (section 6.3). The results shown in this chapter were acquired on MIM capacitors including 9 nm Si:HfO$_2$ films.

6.1 Field cycling effect

The field cycling effect also known as "wake up" behaviour is associated with a recovery of initially pinched antiferroelectric-like hysteresis loops under alternating electrical stress in ferroelectric materials [56], [209], [232], [233]. Several origins of this phenomenon were proposed [56], [209], [234]: (1) space charge accumulation at the grain boundaries, (2) pinning of domain walls by defects due to electric or/and elastic interactions and (3) alignment of defect dipoles along the existing polarisation directions. Therefore, in pristine non-cycled or aged samples certain domain orientations are more favourable, whereas domain wall movements can be simultaneously impeded. A pinching of a polarisation hysteresis appears, if multiple domains with antiparallel polarisation directions are stabilised [71]. Cycling with an alternating electrical field rearranges defects and charges within the material, releasing the domain walls and facilitating polarisation switching of domains in the external field. The electrically cycled cells exhibit commonly higher and more stable remanent polarisation in comparison to pristine non-cycled ones.

The impact of electrical cycling on the ferroelectric properties of 9 nm Si:HfO$_2$ layers was studied for two different film compositions, the first with initial ferroelectric behaviour (4.4 cat% Si) and the second with initial antiferroelectric-like behaviour (5.6 cat% Si). P-V and C-V characteristics were measured after varying number of stress cycles. The stressing was performed using triangular pulses of alternating polarity with amplitude of 3.5 V and frequency of 1 kHz (inset Figure 6.1 (a)). Figure 6.1 shows the evolution of P-V characteristics and corresponding transient current response with increasing number of cycles for ferroelectric films (4.4 cat% Si). The initially slightly pinched hysteresis loops recovered after exposure to the alternating electrical stress. With increasing number of cycles an opening of the hysteresis loop (Figure 6.1 (a)) was observed simultaneously to the alteration of the displacement current characteristics (Figure 6.1 (b)). Fresh non-cycled cells revealed three switching peaks, indicating distribution in the coercive field value for different domains, which also explains a flat slope of the initial hysteresis loop. As argued in [235] the presence of several distinct switching peaks originates from the difference in the local defect concentration. With progressive cycling the slope of the polarisation hysteresis became steeper, which means that the majority of the ferroelectric domains switched almost simultaneously. It was also confirmed by merging of initial three switching peaks into one during cycling. A similar behaviour of displacement current under electrical stress was also reported in [209] and was attributed to the decrease in the internal bias as a consequence of defect dipoles rearrangement. Furthermore, the behaviour of the remanent polarisation (P_R) and polarisation loss (ΔP_R) within 1 s delay during cycling was analysed for samples exposed to different annealing conditions (Figure 6.2). An increase in P_R and more importantly a simultaneous decrease in ΔP_R were detected with increasing number of stress cycles (in particular for anneals at 800 and 1000 ºC). The electrical stress, thus, not only increased the number of switchable ferroelectric domains but also their stability, preventing back switching after removal of external bias. Therefore, not only higher polarisation values can be achieved upon electrical cycling, but also the retention properties of the material can be significantly improved. For 1000 °C anneal, for example, the pristine non-cycled samples exhibited P_R of 22 μC/cm^2. In a long term, however, less than 50 % of this polarisation can be used, taking in account ΔP_R of 11 μC/cm^2 within 1 s. After 1000 cycles P_R rose by 10 %, whereas ΔP_R simultaneously reduced to 3 μC/cm^2, which amounted to 14 % of the P_R-value. "Wake up" behaviour was also reflected in the *C-V* characteristics, which altered with cycling (Figure 6.3). Initial *C-V* curves exhibited double butterfly characteristics with one main and one secondary peak for each sweep direction. Butterfly-shaped characteristics, typical for ferroelectric materials [72], evolved after 1000 times cycling. Moreover, an increase in capacitance value was detected, evidencing release of the domain walls during cycling. Reversible motion of released domain walls in response to an AC signal provided additional contribution to the measured capacitance.

In the case of Si:HfO$_2$ sample with antiferroelectric-like behaviour (5.6 cat% Si) the electrical stress had a negligible impact on the polarisation hysteresis as well as on the transient current response (Figure 6.4). A slight increase in the remanent polarisation with increasing cycling number was detected. The polarisation loss, however, also grew. As a result, the polarisation after 1 s (P_{R_rel}) remained negligibly small. Thus, samples with strongly pinched hysteresis loops (with evident antiferroelectric-like behaviour) cannot be recovered by applying electrical stress. Possible root causes of a cycling behaviour different to the films with 4.4 cat% Si are: (1) material properties different from ferroelectric, where pinched hysteresis arose primarily due to the field induced transition into a ferroelectric phase and not due to domain clamping by defects or (2) a significantly higher concentration of

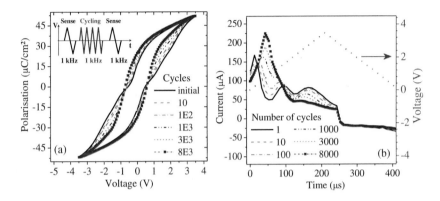

Figure 6.1 Field cycling effect for 9 nm thick ferroelectric Si:HfO$_2$ films with 4.4 cat% Si annealed at 800 °C for 20 s. Evolution of (a) *P-V* characteristics and (b) transient current response with increasing number of cycles. Cycling and measurements were performed using triangular pulses with amplitude of 3.5 V at 1 kHz frequency (inset).

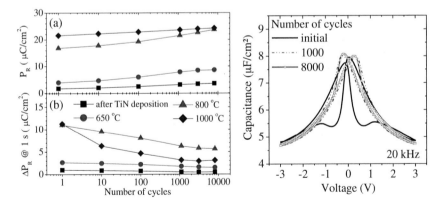

Figure 6.2 Field cycling effect for 9 nm Si:HfO$_2$ films with 4.4 cat% Si depending on the annealing conditions. (a) Dynamic P_R and (b) P_R loss (ΔP_R) after 1 s delay versus number of cycles.

Figure 6.3 Evolution of *C-V* characteristic with increasing number of cycles for 9 nm Si:HfO$_2$ films with 4.4 cat% Si annealed at 800 °C.

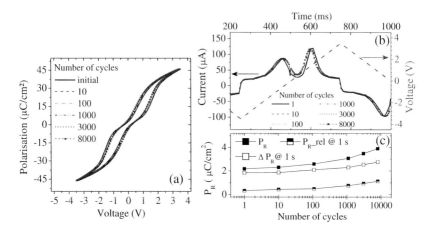

Figure 6.4 Field cycling effect for 9 nm Si:HfO$_2$ films with 5.6 cat% Si annealed at 1000 $^\circ$C. Evolution of (a) *P-V* characteristics and (b) transient current response with increasing number of cycles. (c) Impact of cycling on P_R, relaxed P_{R_rel} and ΔP_R loss after 1 s. using triangular pulses with amplitude of 3.5 V at 1 kHz frequency (inset Figure 6.1 (a)).

defect dipoles induced with additional Si doping, which stiffened the existing domain structure and made it insensitive to the field cycling. The first explanation was supported by the stabilisation of the non-ferroelectric tetragonal phase observed for Si contents higher than 4.4 cat% (Figure 5.13 (a)). The second mechanism, however, cannot be completely excluded.

6.2   Switching kinetics

The switching behaviour of 9 nm thick ferroelectric Si:HfO$_2$ films (4.4 cat% Si) annealed at 1000 °C was studied using PUND technique (chapter 3.2.2). The pulsed characteristics obtained with this method (Figure 6.5 (a)) are more relevant for the memory applications [169]. Taking into account the field cycling effect (section 6.1), the test capacitors were cycled 1000 times with triangular pulses of 3.5 V at 1 kHz before performing PUND measurements. Pulse sequence used during the PUND testing is shown in Figure 6.5 (b). Polarisation switched during the writing pulse ($2P_R$) was monitored depending on the pulse width (t_{WRITE}) and amplitude (V_{WRITE}). Before each writing pulse a cell was set into an opposite polarisation state by applying an initialisation pulse of 3.5 V for 250 μs. The read out was performed with two consecutive pulses (3.5 V / 250 μs each) and read delay of 1 s. The switching times detected for the studied Si:HfO$_2$ samples lay in the nanosecond range and were comparable to those reported for perovskite-type ferroelectric thin films [65], [66], [67].

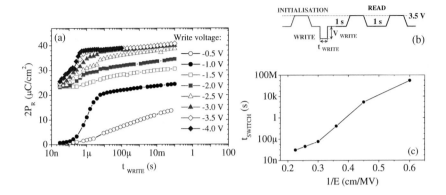

Figure 6.5 Polarisation switching behaviour of 9 nm ferroelectric Si:HfO$_2$ films: (a) Switched polarisation (2P_R) as a function of the writing pulse width for varying pulse amplitudes. (b) PUND experimental pulse sequence. (c) Switching time, time required for complete polarisation switching, versus reversed electric field applied.

For writing voltages higher or equal to – 2 V polarisation reversal was observed already after 30 ns pulses, the shortest pulses available with the used set-up. A saturation of the 2P_R-value was detected at around 40 μC/cm^2, which is in good agreement with the P_R = 24 μC/cm^2 extracted from P-V measurements (Figure 5.5 (a)). The shortest time for complete polarisation reversal was 400 ns achieved for – 4 V. Lower amplitudes of the writing pulse resulted in longer switching times. Moreover, change in the writing voltage caused a change in the slope of the switching curves. In the classical Kolmogorov-Avrami-Ishibashi switching theory [59], [60], however, the switching curves are expected to shift along the time axis for varying switching voltages without any change in their slope [62], [58]. Therefore, the polarisation reversal in Si:HfO$_2$ films proceeded more likely in accordance with the nucleation-limited-switching (NLS) model [61], rather than the classical Kolmogorov-Avrami-Ishibashi switching theory [59], [60]. An essential requirement of the NLS model is the presence of regions with independent switching dynamics [61], which was fully justified for studied Si:HfO$_2$ films due to their polycrystalline structure (Figure 5.2). The grain boundaries in polycrystalline films served as hindrance for domain wall propagation and, thus, infinite domain expansion, resulting in independently switching regions. Nucleation-limited switching kinetics was also confirmed for polycrystalline thin PZT films in [64]. From the experimental switching characteristics of Si:HfO$_2$ films (Figure 6.5 (a)) the switching time, the time of a complete polarisation reversal, was extracted dependent on the writing pulse amplitude. If plotted versus a reversed value of the corresponding applied electric field, the switching time showed an exponential dependence (Figure 6.5 (c)). This corresponded surprisingly well with an empirical law obtained by Scott et al. for submicron PZT films in [60] based on the Kolmogorov-Avrami-Ishibashi model:

$$t_{SWITCH} = t_0 \exp\left(E_\alpha \tau / E\right), \tag{6.1}$$

where t_0 is a constant, E_α – the switching activation field and $\tau = (T_C - T)/T_C$ – the reduced temperature with T_C standing for the Curie temperature. By fitting the experimental curve for Si:HfO$_2$ films and assuming the T_C-value[4] of 200 °C, E_α around 100 MV/cm was obtained. This unreasonably high E_α-value indicated again, that the classical switching theory is not applicable for studied films.

6.3 Fatigue behaviour

This chapter focuses on the fatigue behaviour of Si:HfO$_2$ ferroelectrics, a potential to withstand continuous polarisation reversal. Fatigue characterisation of 9 nm ferroelectric Si:HfO$_2$ films (4.4 cat % Si) annealed at 1000 °C for 1 s was performed. Symmetrical rectangular pulses of alternating polarity were applied to emulate the polarisation switching (inset Figure 6.6 (a)). The remaining switchable polarisation was sensed in between by interrupting fatigue signal and carrying out *P-V* measurements. Here, a constant frequency of 1 kHz and amplitude similar to the stress signal were used. The amplitude and frequency of the stress signal were varied in order to study the influence of these parameters on the fatigue properties.

The stress frequency turned out to affect strongly the number of switching pulses, which devices could withstand before hard breakdown (Figure 6.6 (a)). At the lowest frequency of 10 kHz the switching ability was restricted to 10^4 switching cycles. At the highest frequency of 1 MHz, on the other hand, no breakdown was observed even after 10^9 switching cycles. A similar enhancement of the time to breakdown with increasing frequency was previously observed for high-*k* dielectrics [236] and high-*k* dielectric stacks [237] and was attributed to the trapping effects within the high-*k* material. In addition, breakdown behaviour can be driven by ferroelectric switching as proposed in [238]. Here, heating up of the atoms, moving during polarisation switching, and the resulting bond breakage were held responsible for the breakdown in ferroelectrics. In our case, the amount of switched polarisation differed depending on stress frequency, since the latter determines the width

---

[4] The exact value of the Curie temperature for Si:HfO$_2$ films is still not know. The ferroelectric properties remain at least up to 200 °C as shown by Schröder et al. in [131]. The data for higher temperatures are, however, missing. Therefore, the T_C–value of 200 °C was used for a rough estimation of E_α.

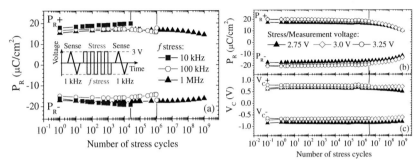

Figure 6.6 Fatigue characteristics of 9 nm ferroelectric Si:HfO$_2$ films: (a) Impact of stress frequency at a constant stress voltage of 3 V. Inset shows experimental pulse sequence. Impact of stress voltage on the alteration in (b) the remanent polarisation and (c) coercive voltages with cycling at 1 MHz stress frequency.

of the polarising pulse (50 μs – for the lowest frequency of 10 kHz and 500 ns – for the highest frequency of 1 MHz). Higher polarisation values were achieved for longer switching pulses/lower frequencies as can be seen from Figure 6.5 (a). Therefore, faster breakdown at lower frequencies according to this theory can be explained by a higher number of atoms moving during cycling and, thus, increased probability of bond breakage. Faster breakdown for higher switched polarisation can, however, also be explained in context of the classical breakdown theory of dielectrics. Here, electric field is known to accelerate the degradation rate [239]. Increase in polarisation enhances electric field within the film during polarisation back switching, which results in earlier breakdown.

Furthermore, the impact of stress voltage on the fatigue characteristics was studied (Figure 6.6 (b) and (c)). The measurements were performed at 1 MHz frequency, since it was comparable to the operation conditions of the Si:HfO$_2$-based FeFET memory cells, which, as will be shown in chapter 7.2, operated in MHz frequency regime with program / erase pulses of 10 – 100 ns (Figure 7.4). For two lowest stress voltages 2.75 and 3 V decrease in the remanent polarisation started after 10^6 cycles (Figure 6.6 (b)), whereas the coercive voltage was almost unaffected by cycling up to 10^9 cycles (Figure 6.6 (c)). This is similar to the fatigue behaviour of PZT films, where cycling predominantly affects P_R-values [78]. At the highest stress voltage of 3.25 V, P_R as well as V_C remained almost constant up to hard breakdown at 10^6 cycles. Therefore, the fatigue properties of Si:HfO$_2$ films were comparable to those of PZT ferroelectrics combined with Pt electrodes, where onset of the polarisation degradation was reported between 10^4 – 10^7 switching cycles [78]. PZT films with oxide electrodes exhibit commonly superior fatigue properties (10^9-10^{12} cycles) [50], [240], whereas SBT films are virtually fatigue free with endurance better than 10^{12} cycles [45], [241]. For the

FeFET-type memories the behaviour of the coercive field with cycling is of greater importance than of the remanent polarisation, since the memory window of the FeFET cell is it predominantly determined by the coercive field and is only weakly affected by the remanent polarisation [107] (equation (2.3)). Hence, in respect to FeFET memory applications Si:HfO$_2$ ferroelectrics demonstrated promising cycling properties (10^9 switching cycles), if operated at moderate voltages and MHz frequencies.

In order to gain better insight into the decrease of the remanent polarisation observed after 10^6 switching cycles, polarisation hysteresis (Figure 6.7 (a)) and transient current characteristics (Figure 6.7 (b)) were analysed at different stress levels (initial unstressed cells, after 10^7 and 10^9 cycles). A tilt of the polarisation hysteresis upon cycling (Figure 6.7 (a)), typical for fatigue phenomenon, was accompanied by change in the intensity and position of the switching peaks (Figure 6.7 (b)). This evidenced an alteration in the distribution

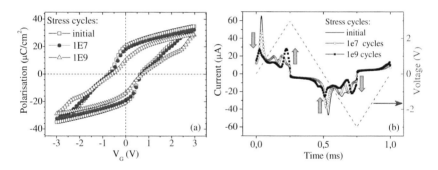

Figure 6.7 Alteration of (a) *P-V* characteristics and (b) transient current response with increasing number of stress cycles during fatigue test at stress voltage of 3 V and frequency of 1 MHz.

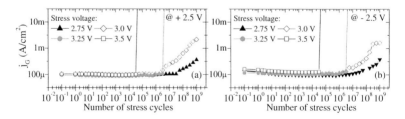

Figure 6.8 Alteration of gate leakage at (a) *+ 2.5 V* and (b) *– 2.5 V* with increasing number of stress cycles during fatigue test at stress voltage of 3 V and frequency of 1 MHz.

of the switching fields of the ferroelectric domains. With progressive cycling the amount of domains requiring high switching fields increased at the expense of domains, which initially switched at low electric fields. Therefore, the fatigue behaviour of Si:HfO$_2$ ferroelectrics can be explained similar to the conventional ferroelectric materials by modification of ferroelectric domains switching capability during cycling [53], [78] – [82]. This was attributed to either domain walls pinning by mobile charged defects [82] – [84] or inhibition of growth of domain nuclei with opposite polarity [78] – [80]. Increase in the leakage current for both polarities after 10^6 switching cycles (Figure 6.8) indicated generation of new defect in Si:HfO$_2$ films during cycling. These generated defects could be responsible for the impairment of the ferroelectric domain switching and, as a result, decrease in the remanent polarisation (Figure 6.6 (b)). For perovskite-type ferroelectrics two main microscopic origins of fatigue were proposed – oxygen vacancies [79], [85], [86], [89], generated and redistributed within the ferroelectric layer under electrical stress, or free charges injected from electrodes [78], [82], [87], [88]. Both mechanisms are possible in Si:HfO$_2$-based ferroelectrics and can potentially influence their fatigue behaviour. Oxygen vacancies are a well-known defect type in HfO$_2$ [144], [161], [160]. A distribution of oxygen vacancies within the film and their concentration at the electrode interface can be affected by an external voltage. This effect is utilised in HfO$_2$-based resistive memories [242]. The charge trapping phenomena in HfO$_2$ are also frequently discussed in the literature [150], [149]. HfO$_2$-based dielectrics were even employed as a storage layer of charge-trapping memories owing to their trapping ability [155], [156]. The study of the exact microscopic origin of the fatigue properties of Si:HfO$_2$ films will remain, however, out of the scope of the present work.

6.4 Summary

The potential of ferroelectric Si:HfO$_2$ films for ferroelectric memory applications has been studied. The effect of field cycling (chapter 6.1), polarisation switching kinetics (chapter 6.2) and fatigue properties (chapter 6.3) has been analysed in detail.

The switching capability of Si:HfO$_2$ ferroelectric films was found to be comparable to the perovskite-type ferroelectric thin films [65], [66], [67]. Switching times in the nanosecond range at voltages as low as 2 V to 4 V (Figure 6.5) could be demonstrated using a pulsed measurement technique. Therefore, Si:HfO$_2$-based memories can be operated in the MHz frequency regime. The switching kinetics in Si:HfO$_2$ films were better described by the nucleation-limited-switching model [61], rather than the classical Kolmogorov-Avrami-Ishibashi switching theory [59], [60]. This behaviour is typical for polycrystalline thin films, as argued in [61], [64], which was also the case for the Si:HfO$_2$ samples investigated in this work. In comparison to the state-of-the-art floating-gate technology, which requires 15 V – 18 V and write times of 1 μs – 1 s [4], Si:HfO$_2$-based memories can provide a significant advantage in terms of operation voltage and programming speed.

Fatigue properties of Si:HfO$_2$ films has been studied depending on frequency and voltage amplitude (chapter 6.3). A MHz frequencies and moderate voltages, Si:HfO$_2$ films demonstrated fatigue characteristics comparable to PZT ferroelectrics combined with Pt electrodes, where onset of the fatigue was reported at between 10^4 and 10^7 switching cycles [78], [50]. In the Si:HfO$_2$ films a decrease in the remnant polarization was detected after 10^6 switching cycles (Figure 6.6 (b)), whereas the coercive voltage was almost unaffected at up to 10^9 cycles (Figure 6.6 (c)). In contrast to the perovskite-type ferroelectrics, dielectric breakdown was found to be an additional factor limiting the cycling capability of Si:HfO$_2$ films. This is a consequence of the high coercive fields of the Si:HfO$_2$ ferroelectrics (~1 MV/cm), bringing the device operation close to material breakdown conditions. The breakdown issue was shown to become more severe at low frequencies and high switching voltages (Figure 6.6). However, at MHz frequencies, corresponding to the operation regime of the FeFET memory cells (chapter 7.2), and moderate voltages, cycling capability of 10^9 cycles could be achieved. Moreover, the reduction of the remanent polarisation in Si:HfO$_2$ with cycling could be attributed to an increase in the number of domains inhibited in switching (Figure 6.7 (b)). The suppression of the switching process during cycling goes hand in hand with generation of new defects, which was confirmed by leakage current measurements (Figure 6.8). An improvement of the fatigue stability of Si:HfO$_2$ may be obtained by introduction of a new electrode materials, as in case of PZT ferroelectrics [50], [240]. Oxygen vacancies thought to be responsible for the fatigue degradation in PZT ferroelectrics [79], [85], [89]. They are also a well-known defect type in HfO$_2$-based dielectrics [144], [161], [160] and can be a possible cause of fatigue in HfO$_2$-based ferroelectrics. Therefore, conductive oxides such as RuO$_2$ and IrO$_2$ are the electrode materials of choice due to their ability to reduce the concentration of oxygen vacancies in the ferroelectric films. The implementation of oxide electrodes for PZT ferroelectrics led to an increased cycling capability of $10^9 - 10^{12}$ cycles [240], [243], [244]. For FeFET applications the evolution of V_C with cycling is of greater importance, since it essentially affects the memory window of the FeFET cell in contrast to P_R, which only has a limited impact on the *MW*. Here, promising characteristics were obtained for Si:HfO$_2$, exhibiting stable V_C values up to 10^9 switching cycles (Figure 6.6 (c)).

The intrinsic defects of Si:HfO$_2$ material were found to impact the operation of Si:HfO$_2$-based ferroelectric devices. The influence of the intrinsic defects became apparent during field cycling experiments (chapter 6.1). Pristine unstressed devices exhibited pinched hysteresis loops (Figure 6.1 (a)) and a wide distribution in the switching fields of the ferroelectric domains (Figure 6.1 (b)). Initially pinched hysteresis loops are not a unique property of Si:HfO$_2$ ferroelectrics. They are also common for conventional ferroelectric materials [232], [233], which is attributed to the interaction of ferroelectric domains with intrinsic defects or defect dipoles [56], [209]. By applying an alternating electrical stress it was possible to open the initially pinched hysteresis loops in ferroelectric Si:HfO$_2$ films (Figure 6.1 (a)) as well as to improve retention properties of the remanent polarisation at the same time (Figure 6.2). A positive effect of the field cycling, also referred to as the "wake up" effect, has also been observed in other ferroelectric materials [209], [232]. Facilitation of switching upon field cycling is explained by a redistribution of the existing defects, resulting in release of stuck domains/pinned domain walls. However, the reverse process of aging is very likely to progress with time [234], where either the defect dipoles reorientate, stabilising the existing domain polarisation, or the defects diffuse to domain walls, pinning them due to electric and/or elastic interactions. Hence, the intrinsic defects of Si:HfO$_2$ should be also expected to affect its retention properties. Moreover, the impact of intrinsic defects on the fatigue behaviour of Si:HfO$_2$ ferroelectrics cannot be ruled out. Although a correlation between fatigue and generation of new traps (chapter 6.3) was found, the contribution of pre-existing defects to the degradation is not yet completely clear.

7 Ferroelectric field effect transistors based on Si:HfO$_2$ films

Si:HfO$_2$-based ferroelectrics exhibit several advantages for application in ferroelectric field effect transistors (FeFETs) in comparison to perovskite-type ferroelectric materials: full CMOS compatibility and a better scaling potential of the gate stack. The latter is assisted by stable ferroelectric properties at film thicknesses in the nanometre range (5 – 30 nm), an order of magnitude higher coercive field ($E_C \sim 1$ MV/cm) (Figure 5.10(c)) in combination with an order of magnitude lower dielectric constant of $\sim$25. The feasibility of the FeFET devices on the basis of Si:HfO$_2$ ferroelectric films as well as their basic operation characteristics were previously demonstrated in [22], [229], [245]. The emphasis of this work lies on a more detailed study of the performance of Si:HfO$_2$-based FeFETs in order to get better understanding of the physical mechanisms affecting their main properties (program and erase characteristics (chapter 7.2), retention (chapter 7.3) and endurance behaviour (chapter 7.4)). Devices with gate lengths of 260 nm were chosen for analyses in order to assure comparability to the state-of-the-art FeFET cells with perovskite-type ferroelectrics [112]. Moreover, a scaling potential of Si:HfO$_2$-based FeFETs down to a contemporary CMOS technology node of 28 nm was investigated in this work for the first time. The impact of scaling on the device memory characteristics will be discussed in chapter 7.5. The influence of Si:HfO$_2$ film composition on the operation of the FeFET cells will be shown in chapter 7.1.

7.1   Effect of the silicon doping

The electrical behaviour of Si:HfO$_2$ films was shown previously to depend strongly on the Si doping level (chapter 5.1.1). MFIS-FETs containing 9 nm thick Si:HfO$_2$ layers with different silicon contents (3.7, 4.4 and 5.7 cat% Si) were fabricated so that the influence of Si doping could be also studied for transistor structures. N-channel MFIS-FETs devices with a gate length (L_G) of 260 nm and a gate width (W_G) of 2 µm were analysed in this and following sections unless mentioned otherwise. I_D-V_G characteristics were measured on unstressed devices (initial) as well as after applying a positive gate pulse of + 6 V for 100 ns and a negative gate pulse of – 6 V for 100 ns (Figure 7.1). The response of the I_D-V_G curves to the applied pulses differed depending on the Si content. Devices with Si:HfO$_2$ layers containing 3.7 and 4.4 cat % Si showed curve shifts opposite to the polarity of the applied gate pulses (negative for a positive gate pulse and positive for a negative gate pulse), which is characteristic for a ferroelectric switching (Figure 2.11). For devices with 5.7 cat% Si, on the other hand, a prevailing charge trapping behaviour was identified. Here, an inverse behaviour was observed: the I_D-V_G characteristics shifted in the direction similar to the polarity of the applied gate pulse. This trend corresponded well with the electrical properties of the MFM capacitors discussed in chapter 5.1.1, where distinct ferroelectric behaviour was detected for samples with 4.4 cat% Si, whereas antiferroelectric-like characteristics appeared for Si contents ≥ 5.6 cat% (Figure 5.1).

A die-to-die distribution of the memory window on 300 mm wafers depending on the Si content was examined (Figure 7.2). A homogeneous *MW* distribution within the wafers was detected. The largest *MW* of about 1.2 V was obtained for 4.4 cat% silicon doping. These samples with the most pronounced ferroelectric properties were used in further studies.

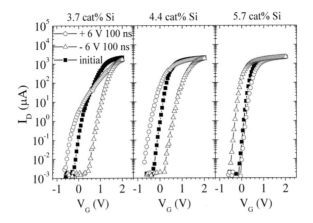

Figure 7.1 I_D-V_G characteristics measured on unstressed devices, after + 6 V/ 100 ns and – 6 V/ 100 ns pulses for MFIS-FETs (L_G = 260 nm, W_G = 2 µm), including Si:HfO$_2$ films with varying Si content. Devices with layers containing 3.7 and 4.4 cat% Si exhibited a predominant ferroelectric behaviour distinguished by a shift of the I_D-V_G curves opposite to the polarity of the applied gate pulses. An inverse behaviour for layers with 5.7 cat% indicated a prevalence of charge trapping.

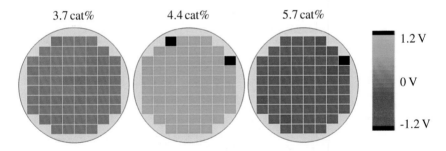

Figure 7.2 Die-to-die memory window distribution of MFIS-FETs (L_G = 260 nm, W_G = 2 µm) on 300 mm wafers with different Si contents of the Si:HfO$_2$ layer. Program and erase were performed using pulses of – 5 V/ 100 ns and + 5 V/ 100ns, respectively.

7.2 Program and erase operation

Writing speed and voltages required for reversal of a memory state during program and erase operation were analysed for MFIS-FETs (L_G = 260 nm and W_G = 2 μm) containing ferroelectric Si:HfO$_2$ films with 4.4 cat% Si. By applying a positive erase (+ 4 V/ 100 ns) and a negative program (− 6 V/ 100 ns) pulse a shift of the I_D-V_G curves in the direction opposite to the polarity of the applied pulses was induced (Figure 7.3 (a)), which confirmed a ferroelectric switching. Furthermore, program and erase characteristics (Figure 7.3 (b, c)) were measured using pulses with varying width (10 ns – 100 μs) and amplitude (2 – 6.5 V). Each point of the shown characteristics corresponds to a memory window after applying a program/erase pulse of given width and amplitude. Prior to each writing pulse (program or erase) a studied FeFET cell was set into a completely erased or programmed state by applying an initialisation pulse of + 4 V/ 100 ns or − 6 V/ 100 ns, respectively. The inset of the Figure 7.3 (c) shows an example of the pulse sequence used for obtaining erase characteristics. A non-zero *MW*, indicating the polarisation switching, was detected already after 10 ns pulses for erase voltages ≥ + 4 V and program voltages above − 5 V. Reduction of the memory window with increasing pulse width was probably caused by charge trapping

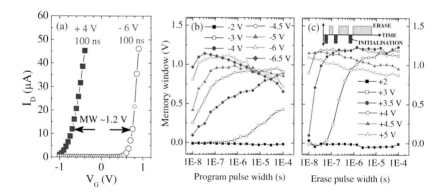

Figure 7.3 Program/erase characteristics of MFIS-FETs containing a Si:HfO$_2$ film with 4.4 cat% Si (L_G = 260 nm, W_G = 2 μm): (a) I_D -V_G characteristics after an erase (+4 V/ 100 ns) and a program pulse (− 6 V/ 100 ns), resulting in a memory window of 1.2 V. Memory window as a function of (b) program and (c) erase pulse width for varying pulse amplitudes.

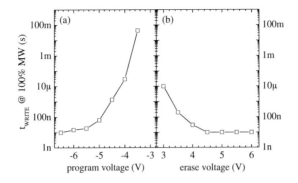

Figure 7.4 Pulse width required to obtain a maximum memory window depending on the pulse amplitude for (a) program and (b) erase operation.

from the transistor channel: electron trapping for positive gate pulses and hole trapping for negative gate pulses. A V_{TH} shift opposite to that induced by ferroelectric switching is characteristic for charge trapping. After some pulse width the switched ferroelectric polarisation must have saturated, whereas the amount of trapped charge continued to increase with increasing pulse width, resulting in degradation of the memory window. Moreover, from the program/erase characteristics the most optimal conditions for read operation can be determined. Disturb-free readout can be presumed in the voltage range between − 2 V and +2 V, where no polarisation reversal (a zero MW) was detected up to 100 µs. Furthermore, a voltage dependence of the pulse width required to obtain a maximum memory window was extracted (Figure 7.4). Similar to the capacitor switching behaviour (Figure 6.5), higher voltage amplitude resulted in shorter switching times for both program and erase operations. A trade-off between fast switching speed and low power operation could be achieved at + 4 V erase and − 5.5 V program voltage, where the maximum memory window of 1.2 V was obtained with pulses of 30 ns. The operation voltages for Si:HfO₂-based MFIS-FET structures were slightly higher in comparison to MFM capacitors (chapter 6.2). The main reason was an additional interfacial 1.2 nm SiON layer integrated into the transistor gate stack, which reduced the effective voltage drop over the ferroelectric layer. Comparable operation voltages were reported (4 − 7 V) for the FeFET cells with perovskite-type ferroelectric SBT films at the same 260 nm node [112]. Their switching speed was, however, inferior to that demonstrated for Si:HfO₂-based devices. Pulses of 100 ms resulted in only half of the MW of the devices studied here.

7.3   Retention behaviour

The non-volatility of a memory cell is defined as its capability to store the data without power supply for at least 10 years [1]. This specification has been a challenge for FeFET type memories with perovskite-type ferroelectric materials until an introduction of high-*k* buffer layers at the Si interface [106], [105], [246]. Promising retention results were reported for Si:HfO$_2$-based ferroelectrics so far. A stable ferroelectric polarisation in MFM capacitors with Si:HfO$_2$ ferroelectric films was demonstrated for 20 hours at 125 °C [247]. Furthermore, Si:HfO$_2$-based FeFET devices were predicted to exhibit a residual memory window after 10 years at room temperature [229], [245]. In this work the impact of temperature as well as programming conditions on the retention characteristics of Si:HfO$_2$-based FeFETs was examined.

The temperature dependence of the retention behaviour was studied on FeFET devices with L_G = 260 nm and W_G = 2 μm at 25, 150 and 210 ºC. The retention measurements were performed using the procedure described in chapter 3.1.3. Figure 7.5 (a) shows the resulting retention characteristics for both "ON" and "OFF" memory states. These were initially set by applying gate pulses with a width of 100 ns and voltages of + 4.5 V and – 6.5 V, respectively. The operation capability of the studied memory cells, in which two distinguishable memory states could be established, was proven up to 210 °C. A linear behaviour on the logarithmic time scale was observed for both memory states at all temperatures. By extrapolating the experimental trends to 10 years a residual memory window was estimated depending on temperature (Figure 7.5 (b)). An increased in temperature caused the acceleration of the time dependent V_{TH} decay for both "ON" and "OFF" states, which resulted in a decrease of the residual *MW*. Nevertheless, the obtained characteristics predicted a residual *MW* of 0.2 V after 10 years even at the highest temperature of 210 °C.

The impact of the program and erase pulse amplitude on the retention behaviour of Si:HfO$_2$-based FeFETs was investigated at 30 °C (Figure 7.6). All pulses exhibited a width of 100 ns. The V_{TH} shift after 10 days for program and erase pulses of varying amplitudes is shown in Figure 7.6 (a). Higher pulse amplitudes were accompanied by a decrease of the V_{TH} shift (ΔV_{TH}) and, hence, improved retention properties. For example, the V_{TH} shift was lowered by almost 90 % by performing the programming at – 7 V instead of – 6 V. The same trend was observed for erase operation. The erase pulse of + 5 V instead of + 4 V not only reduced the V_{TH} shift but also induced a change in the shift direction. In this case the memory window increased slightly with time. Moreover, the V_{TH} value directly after the program/erase pulse also showed a dependence on the amplitudes of the program (Figure 7.6 (b)) and

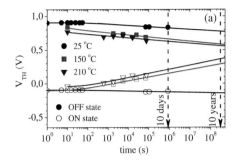
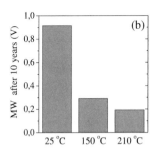

Figure 7.5 (a) Retention characteristics for "OFF" (after – 6.5 V/ 100 ns) and "ON" (after + 4.5 V/ 100 ns) memory states at varying temperatures. (b) Residual memory window after 10 years as a function of temperature. Measurements were performed on MFIS-FET devices with $L_G = 260$ nm and $W_G = 2$ μm.

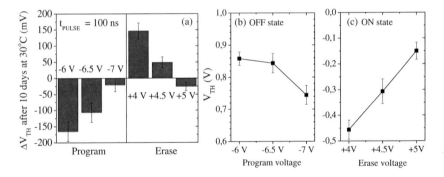

Figure 7.6 (a) Impact of program/erase pulse amplitude on the V_{TH} shift (ΔV_{TH}) after 10 days at 30 °C. V_{TH} value directly after a writing pulse as a function of pulse amplitude for (b) program and (c) erase.

erase pulses (Figure 7.6 (c)). An increase in program and erase voltages degraded the V_{TH} in the corresponding memory state. V_{TH} in the "OFF" state (Figure 7.6 (b)) shifted to more negative voltages, whereas V_{TH} in the "ON" state shifted to more positive voltages. Both effects lead to a reduction of the initial MW. This reduced MW and at the same time improved retention, which were observed for higher program/erase voltages, can be explained in terms of the charge trapping effect. The fact that charge trapping accompanied the ferroelectric switching in Si:HfO$_2$ FeFETs has been detected during the study of their program and erase behaviour (chapter 7.2). These trapped charges caused a partial compensation of the ferroelectric polarisation, which resulted in lowering of the memory window (Figure 7.3 (b, c)). This partial compensation of the polarisation charge has, however, a positive effect, namely reduction of the depolarisation field. The depolarisation field is considered as one of the main driving forces for retention loss in MFIS transistors [110], [248]. Therefore, its reduction due to increased amount of trapped charges at higher program/erase pulse amplitudes caused the observed improvement of the retention behaviour.

7.4 Endurance properties

This chapter deals with the endurance properties of Si:HfO₂-based FeFET devices, an ability of a memory cell to withstand a continuous program/erase operation (chapter 3.1.4). A cycling capability of Si:HfO₂ films implemented into MFM capacitors was discussed in chapter 6.3. At low frequencies (10 – 100 kHz) and high switching fields (>3 MV/cm) the maximum number of switching cycles was limited to $10^4 - 10^6$ by a dielectric breakdown rather than a classical fatigue mechanism. Operation at MHz frequencies and moderate electric fields (2.5 – 3 MV/cm) enabled to extend the cycling capability to 10^9 cycles. A fatigue-free behaviour up to 10^6 cycles and, most importantly for FeFET operation, a negligible change in the coercive fields up to 10^9 cycles was demonstrated.

Endurance testing of Si:HfO₂-based MFIS-FETs was performed using a pulse sequence schematically illustrated in Figure 7.7 (a). Alternating pulses of – 6 V / 100 ns and + 4 V / 100 ns were used to emulate continuous program and erase operation. After a certain number of stress pulses a switching capability of the memory cell was verified by setting it into an "ON"- and "OFF" state and reading out the corresponding V_{TH} values by performing an I_D-V_G sweep. Figure 7.7 (b, c) shows typical endurance characteristics – V_{TH} values for the "ON" and "OFF" memory states as well as the resulting memory window versus number of program/erase cycles. An initial increase in the *MW* observed up to 10^3 cycles correlated with the improvement of the ferroelectric behaviour observed for the MFM capacitors upon field cycling (chapter 6.3). This was attributed to redistribution of defects within the ferroelectric film, leading to unpinning of domain walls and/or frozen ferroelectric domains, which in turn facilitated switching and improved polarisation stability. A rapid *MW* degradation after 10^3 cycles was quite surprising taking into account promising results of the Si:HfO₂-based capacitors (Figure 6.6 (c)). FeFET operation at a MHz frequency, at which capacitor structures exhibited the best cycling capability, was assured by program/erase pulses of 100 ns. A hard breakdown, a cause of early failures in capacitors, can be excluded for transistor structures due to a gradual nature of degradation. Therefore, degradation in Si:HfO₂-based MFIS-FETs proceeded apparently in a different way than in MFM structures. The reduction of the *MW* with progressive cycling was mainly determined by a positive shift of V_{TH} in the "ON" state, whereas the "OFF" state V_{TH} remained almost unaffected (Figure 7.7 (b, c)). Similar asymmetric V_{TH} shift is characteristic for floating-gate type cells,

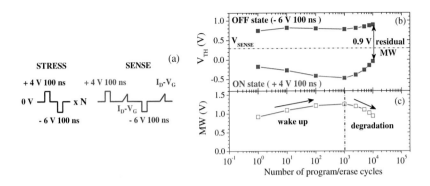

Figure 7.7 Endurance characteristics of Si:HfO$_2$-based MFIS-FETs (L_G = 260 nm and W_G = 2 µm): (a) Experimental pulse sequence for endurance testing; (b) V_{TH} in the "ON" and "OFF" memory states and (c) a corresponding memory window versus number of applied program/erase cycles.

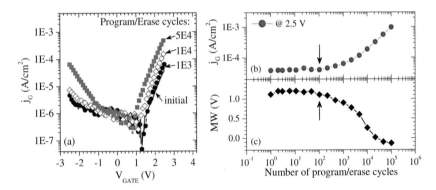

Figure 7.8 Evolution of gate leakage current under continuous program/erase stress: (a) gate current density (j_G) versus gate voltage (V_G) characteristics at different stages of stress testing; Dependence of (b) j_G at 2.5 V and (c) memory window on the number of program/erase cycles.

which is explained in terms of negative oxide charges generated under cycling stress [249]. The endurance capability of studied FeFET devices, demonstrating a residual *MW* of 0.9 V after 10^4 cycles (Figure 7.7 (b, c)) was comparable to that of FG cells [4].

The possibility of the gate stack degradation in Si:HfO$_2$-based FeFET structures during program/erase cycling was examined by simultaneously monitoring the gate leakage current (Figure 7.8). An increase in the gate current correlated well with the onset of the *MW* degradation (Figure 7.8 (b)). Furthermore, gate leakage induced during endurance stress was verified on devices with varying gate areas (Figure 7.9). In both cases for unstressed devices (Figure 7.9 (a)) as well as after 10^5 program/erase cycles (Figure 7.9 (b)) the gate current densities overlay for all devices[5]. Thus, the gate leakage scaled with gate area, which indicated a homogeneous rather than localised degradation of the gate stack area upon cycling stress.

Trap densities within the gate stack were additionally characterised at different stages of the endurance testing in order to identify the main degradation path. A charge pumping (CP) technique (chapter 3.3.1) was used to sense the interface traps as well as traps within the interfacial SiON layer. The bulk traps within the Si:HfO₂ layer, on the other hand, were probed by means of the single-pulse methodology (chapter 3.3.2). Figure 7.10 shows the results of the CP analyses. An increase in the charge pumping current (I_{CP}) (Figure 7.10 (a)) indicated a generation of addition interface traps induced by the cycling stress. From the maximum values of the charge-pumping current the interface trap densities (N_{CP}) was calculated for corresponding number of program/erase cycles (Figure 7.10 (c)) by using the equation (3.3). A strong increase in the interfacial trap density after 10^3 cycles goes along with an increase in the gate leakage current (Figure 7.9). The main contribution to I_{CP} at 1 MHz frequency (Figure 7.10 (a, c)) comes from the traps located directly at Si-SiON interface. The interaction of the substrate charge carriers with deeper traps located within the SiON layer becomes possible at lower frequencies, at which the charge carries have sufficient time to tunnel to the traps and back during the substrate inversion and accumulation periods [193], [194]. The frequency dependence of the trap density was measured at V_{GL} of -1 V for different program/erase stress (Figure 7.10 (b)). An additional contribution from deep traps was seen at low frequencies. It enlarged with continuous program/erase stress, providing evidence that new traps were generated also within the interfacial SiON layer. A quantitative estimation of the trap density in SiON was made at 50 kHz by subtracting the trap density at 2 MHz that corresponded to interface traps (Figure 7.10 (d)). Similarly to interface traps (Figure 7.10 (c)), an apparent increase of the SiON trap density was detected after 10^3 program/erase cycles.

---

[5] The gate current density for the smallest devices was slightly higher in comparison to other devices. This is believed to be a measurement artefact due to very low absolute current values (0.1 pA) close to the resolution limit of the measurement system.

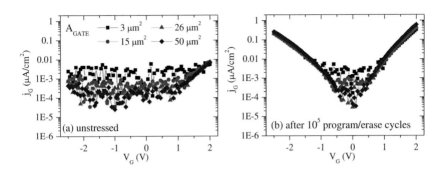

Figure 7.9 Gate area dependence of the gate current density characteristics for (a) unstressed devices and (b) after 10^5 program/erase cycles.

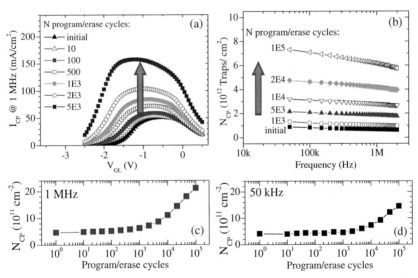

Figure 7.10 Charge pumping characteristics under progressive program/erase cycling: (a) Variable base level charge pumping characteristics – charge pumping current (I_{CP}) versus low level of the gate excitation pulse (V_{GL}) and (b) frequency dependence of the charge-pumping trap density; (c, d) Interface trap density, extracted from the measurements at 1 MHz and 50 kHz, respectively, as a function of program/erase cycles.

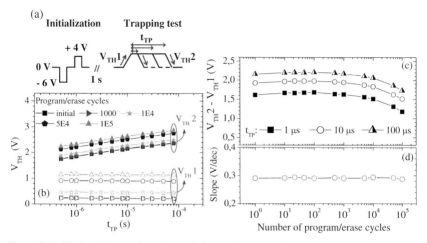

Figure 7.11 Single-pulse trapping characteristics under progressive program/erase cycling [250]: (a) Experimental gate pulse sequence; (b) V_{TH} measured on the rising ($V_{TH}1$) and falling ($V_{TH}2$) pulse edges for varying trapping pulse width (t_{TP}) depending on the number of program/erase cycles; Evolution of (c) the trapping window $\Delta V_{TH} = (V_{TH}2\text{-}V_{TH}1)$ for 1 μs, 10 μs, 100 μs trapping pulses and (d) the slope of ΔV_{TH} (t_{PULSE}) characteristic with increasing number of endurance cycles.

A single-pulse charge-trapping technique (chapter 3.3.2), which allows capturing fast transient trapping, was applied to monitor the evolution of trap density within the Si:HfO₂ layer. HfO₂-based materials are known to exhibit a high density of intrinsic defects (10^{12} – 10^{14} cm⁻²) [141], [142], which can serve either as electron and/or hole traps [141], [145]. The single-pulse measurements were performed on the studied MFIS-FET devices and revealed a high density of electron traps in the Si:HfO₂ layer (see chapter 8.1). During the standard DC measurements, which were shown in chapter 7.2, positive gate pulses were seen to induce a negative V_{TH} shift due to the polarisation switching. A contrary behaviour was observed by the single-pulse tests. The I_D-V_G characteristics in this approach were recorded directly at the rising and falling edges of the excitation pulses, so that the time delay between stressing and sensing was practically eliminated. A positive instead of a negative V_{TH} shift was detected immediately after a positive gate pulse even for unstressed devices (Figure 7.11 (b)). This indicated a strong electron trapping within the gate dielectric, which prevailed over the positive ferroelectric polarisation charge. The latter became dominating only after several microseconds, when detrapping processes have set in (Figure 8.4) and a negative V_{TH} shift was measured. This scenario corresponded to the standard DC measurements. Trapping behaviour onto the Si:HfO₂ traps was examined after different number of program/erase cycles by means of single-pulse technique (Figure 7.11). A pulse sequence, which is schematically shown in Figure 7.11 (a), was used. The V_{TH} shift between the rising and the falling edges of the trapping pulse was measured as a function of pulse width (t_{TP}) for trapping voltage of 3.5 V. Before each trapping pulse an identical initial state was re-established with a combination of a negative (– 6 V/ 200 ns) and a positive (+ 4 V/ 200 ns) pulse that was followed by a delay of 1 s. The resulting trapping characteristics are depicted in Figure 7.11 (b). The trapping window (ΔV_{TH}) defined as a difference between $V_{TH}2$ and $V_{TH}1$ was determined for 1, 10 and 100 µs trapping pulses and plotted versus the number of program/erase cycles (Figure 7.11 (c)). A negligible change of ΔV_{TH} was visible up to 10^4 cycles. For higher cycling numbers the trapping window even started to decrease. This was predominantly determined by the shift of $V_{TH}1$ to higher positive voltages, reflecting the degradation of the "ON" memory state (Figure 7.7 (b)). On the other hand, $V_{TH}2$ remained almost unaffected by cycling. A decrease of the trapping window with cycling indicated a build-up of permanent negative charge within the dielectric stack. This resulted in a modified field distribution within the gate stack and a reduction of the injection current during trapping pulses. Possible origins of this permanent negative charge are fixed charges generated during cycling or accumulation of electrons stuck on the deep traps. Moreover, the slope of $\Delta V_{TH}(t_{TP})$ characteristic, which is proportional to the trap density in material [251], remained independent on the number of program/erase operations (Figure 7.11 (d)). Therefore, it can be deduced that only negligible generation of new bulk traps within the Si:HfO$_2$ layer occurred during endurance stress. Thus, the degradation of the interfacial SiON layer rather than the Si:HfO$_2$ film was mainly responsible for the observed increase of the gate leakage current (Figure 7.8) [250]. A prevailing degradation of the interfacial layer was also ascertained for standard high-k transistor gate stacks exposed to positive or negative bias stress [150], [190], [252]. Alternating stress was reported to accelerate this degradation [253], [254], which was assigned to the wear-out of the interfacial layer due to continuous back and forth tunneling of charges [255].

Furthermore, the impacts of bipolar switching pulses and unipolar stress pulses, referred to as dynamic imprint, on the endurance characteristics were compared (Figure 7.12). The switching capability was verified by setting the device into the "ON" and "OFF" state and reading out the corresponding V_{TH} values, as shown in Figure 7.12 (a), after certain number of unipolar/bipolar stress pulses. The experiment was performed for two sets of program/erase pulses: + 4 V/ – 6 V and + 5 V/ – 7 V. The width of all pulses was equal and amounted to 100 ns. The same trend was observed for both stress conditions (Figure 7.12 (b, c)). The unipolar stress pulses caused only slight V_{TH} shifts of both memory states. Since the induced V_{TH} shifts exhibited signs similar to the polarity of the applied gate pulses, electron/hole injection from the substrate must have predominantly determined the effect of positive/negative unipolar pulses. An inverse behaviour is expected in case of a dynamic imprint associated with stabilisation of stored polarisation states [256]. The bipolar pulses, on the other hand, resulted in a pronounced *MW* degradation. From that it can be deduced, that the polarisation switching itself, similar to classical fatigue behaviour, or mechanisms coupled with alternating pulses aggravate the endurance degradation in transistor structures. The reliability issues of the transistor gate stack rather than the ferroelectric layer were held responsible for the endurance properties of Si:HfO₂-based FeFET devices in [250]. This conclusion was based on the correlation, which was found between the degradation of the interfacial layer and reduction of the memory window. Additional confirmation for this correlation was obtained in this work. The gate leakage current and trap density of the interfacial SiON layer were tested for unipolar and bipolar stress conditions (Figure 7.13). Both these parameters showed an increase only in case of the bipolar pulses, which was similar to the behaviour of the memory window (Figure 7.12). Therefore, there is a strong link between the damage of the interfacial layer, caused by cycling, and deterioration of the FeFET memory operation. The gate stack structure of the studied ferroelectric transistors was similar to an ordinary high-*k* metal gate stack. A thicker and crystalline high-*k* layer exhibited, however, additionally ferroelectric properties. Hence, similar degradation mechanisms under electrical stress can be expected in both stacks. The interfacial layer has been already identified in the main reliability concern in conventional high-*k* metal gate stacks [150], [190], [252]. Here, a continuous charge transport through the interfacial layer results in generates of new traps under alternating bias stress [253], [254], which impairs the insulating properties of the interfacial layer and, thus, the entire gate stack. Figure 7.14 shows energy band diagrams for a p-Si/SiO₂/HfO₂/TiN gate stack, with a structure similar to that of transistors studied in this works, under a positive (+ 4 V) and negative (− 6 V) gate voltage for two cases: (1) paraelectric HfO₂ layer ($P_R = 0$ μC/cm²) and (2) ferroelectric HfO₂ layer ($P_R = 10$ μC/cm²). The energy band diagrams were calculated by means of the multi-dielectric-energy-band-diagram-program [257], [258] using stack parameters, which are listed in Table 7.1.

Table 7.1 Parameter used for the calculation of the energy band diagram of a Si/SiO₂/HfO₂/TiN stack [258].

Parameter	Si	SiO₂	HfO₂	TiN
N_A, cm⁻³	10^{16}			
d, nm		1.2	10	10
ε	11.7	3.9	25	
Eg, eV	1.21	8.9	5.7	
χ^e, eV	4.05	0.95	2.65	
Φ_M, eV				4.45

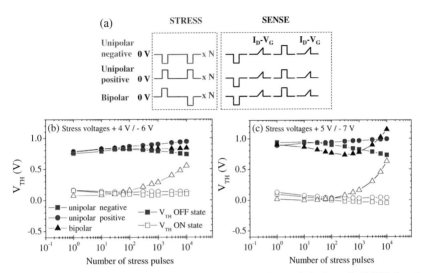

Figure 7.12 Effect of unipolar and bipolar stress pulses on the endurance behaviour of a Si:HfO₂-based FeFETs: (a) Experimental gate pulse sequence for three different stress conditions; Endurance characteristics depending on the applied stress type for two sets of program/erase conditions (a) – 6 V / + 4 V and (b) – 7 V / + 5 V. All pulses used for stress as well as sense operations had a width of 100 ns.

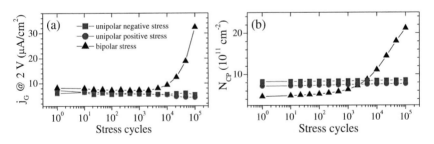

Figure 7.13 Effect of unipolar and bipolar stress pulses on (a) the gate leakage current density and (b) the interface trap density. Program/erase stress conditions were – 6 V / + 4 V with 100 ns pulse width.

In comparison to the high-k gate stack, which includes a non-ferroelectric HfO₂ layer, the charge injection into the ferroelectric transistor is further enhanced. The ferroelectric polarisation charge induces internal fields, which facilitate charge injection from the channel. The positive polarisation established by positive gate voltages assists the electron injection. These electrons (e⁻) injected into the HfO₂ layer are trapped in to the trap states (E_T).

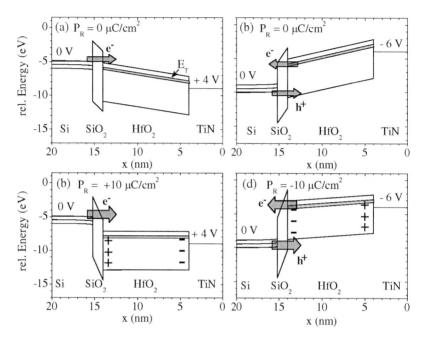

Figure 7.14 Energy band diagrams of the p-Si/SiO₂/HfO₂/TiN stack under positive (+ 4 V) and negative (– 6 V) gate voltage, representing erase and program operation of the Si:HfO₂-based FeFET memory cell: (a, b) Case of a paraelectric HfO₂ ($P_R = 0$ μC/cm²) and (c, d) ferroelectric HfO₂ ($P_R = 10$ μC/cm²). The directions of the electron (e⁻) and hole (h⁺) flows are shown with arrows, which thickness represents the tunnelling probability.

The negative polarisation induced by negative gate voltages facilitates hole injection (h⁺) into the HfO₂ and back tunneling of electrons from the trap states at the same time. Therefore, steady transfer of charges through the interfacial layer is inevitable in the ferroelectric stack under alternating program/erase pulses. The resulting degradation of the insulating properties of the interfacial layer may aggravate the memory operation of the FeFET [250]. Free charges can be easily injected from the channel into the ferroelectric layer, compensating its polarisation that eventually degrades the ferroelectric memory window. Charge injection was one of the issues identified in the first perovskite-based FeFET devices [101], [102], for which the ferroelectric layer was fabricated directly on the semiconductor substrate. Other causes of the memory window degradation with cycling cannot, however, be completely excluded. One of such causes is a build-in of negative charge within the gate stack due to generation of negatively charged defects within the interfacial layer or charging of the pre-existing traps of the interfacial SiON or Si:HfO₂ layers. It has been demonstrated using an energy band diagram (Figure 7.14) that electron injection at positive gate voltages is enhanced in the ferroelectric stack. If these trapped electrons cannot be completely detrapped during negative voltage pulses, they accumulate in the gate stack with increasing number of endurance cycles. This alters the distribution of the electric field in the stack during program and erase pulses and leads to an asymmetric degradation of memory states similar to the behaviour of floating-gate type cells [249]. A build-in of the negative charge can also explain a decrease in the trapping window under endurance stress (Figure 7.11 (c)). In order to separate the effects of the degrading interfacial layer and accumulation of the negative charge on the memory operation of the studied FeFETs, further studies are required.

7.5 Impact of scaling on the device performance

An essential requirement for a successful industrial implementation of a new memory concept is the scaling capability of a single memory cell. This requirement is determined by a continuous demand for the higher data storage densities at lower costs. FeFET devices based on the perovskite ferroelectrics are unlikely to scale below 50 nm due to a physical gate stack height of several hundred nanometres (200 – 500 nm) [13]. The most aggressively scaled devices reported in the literature so far achieved a gate length of 260 nm [112]. HfO_2-based ferroelectrics, on the other hand, exhibit a potential to overcome the limitations of conventional ferroelectric materials due to significantly higher coercive field strength E_C of ~1 MV/cm (for PZT or SBT ~ 50 kV/cm) in combination with a lower dielectric constant of ~25 (for PZT or SBT ~ 200 – 300). At reduced ferroelectric thickness these material properties enable to avoid high depolarization fields and compensate the memory window loss. The gate stack height can be lowered to several nanometres, which provides gate stack aspect ratios more suitable for scaling. It has been recently demonstrated, that FeFETs based on $Si:HfO_2$ ferroelectric thin films can be fabricated at a state-of-the-art 28 nm technology node [24]. This finally has closed the scaling gap between the ferroelectric and CMOS logic transistors. In this chapter the memory properties of the $Si:HfO_2$-based MFIS-FeFET devices scaled down to 28 nm gate length are studied. The impact of device scaling on the key memory characteristics such as program/erase operation, endurance behaviour and retention properties is discussed.

The operation capability of the $Si:HfO_2$ MFIS-FETs was proven down to the smallest device size with gate width of 40 nm and gate length of 28 nm. Figure 7.15 shows an example of I_D-V_G characteristics for these devices in the programmed (after – 5 V/100 ns pulse) and erased (after + 5 V/100 ns pulse) states with a resulting memory window (*MW*) of 0.8 V. The ferroelectric switching effect is confirmed by the shift of the I_D-V_G curves opposite to the polarity of the applied voltage.

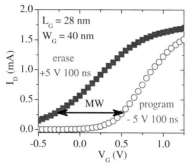

Figure 7.15. Memory operation of the Si:HfO₂-based MFIS-FET device with L_G = 28 nm and W_G = 40 nm. I_D-V_G characteristics after a program (after − 5 V 100 ns) and an erase (after + 5 V 100 ns) pulse. Shifts opposite to the polarity of the applied gate pulse confirm a ferroelectric switching behaviour.

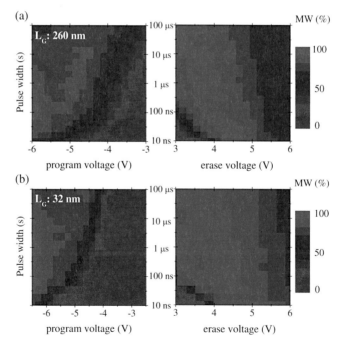

Figure 7.16 Impact of gate scaling on the program/erase operation of Si:HfO₂-based MFIS-FeFETs. The memory window (*MW*) as a function of pulse width and amplitude for negative (program) and positive (erase) voltages measured for two device types: (a) device 2 (L_G = 260 nm, W_G = 2 μm) and (b) device 4 (L_G = 32 nm, W_G = 1 μm).

The impact of gate length scaling on the writing speed and operation voltages, which are required for program/erase, was examined. Program and erase characteristics were measured using pulses with varying width (10 ns – 100 μs) and amplitude (3 – 6 V) for devices with different gate lengths and comparable gate width (devices 1 – 4 in Table 2). No consistent correlation between the maximum achievable MW (0.8 V – 1.4 V) and transistor gate length was found. Figure 7.16 shows two examples of the resulting switching matrixes for devices with gate lengths of 260 nm (device 2) and 32 nm (device 4). Each point of the program/erase matrix corresponds to the MW after applying a pulse of given width and amplitude. Prior to each writing pulse (program or erase) the identical cell state (completely erased or programmed) was restored by applying an initialization pulse of the opposite polarity in respect to the writing pulse (Figure 3.2 (a)). The amplitude and width of the initialization pulses were chosen from the previous studies. Pulses of + 4 V/100 ns and – 6 V/100 ns were used for setting completely erased and programmed state, respectively. For both device types a non-zero MW, which indicated ferroelectric switching, was demonstrated for program and pulses as short as 10 ns (Figure 7.16). With decreasing gate length, however, the voltages required to program cells at the same speed increased. A clear shift in the onset of switching to higher program voltages is seen for the 32 nm devices. The same measurement procedure was carried out for other devices (Table 2). The pulse width (t_{WRITE}), required to achieve the maximum MW, was determined dependent on the program/erase pulse amplitude (Figure 7.17 (a, b)). A decrease in the gate length resulted in a shift of the program characteristics to higher voltages (Figure 7.17 (a)). The erase characteristics shifted simultaneously to lower voltages (Figure 7.17 (b)) with the exception of the curve for the 32 nm device. In order to elucidate the observed behaviour, the I_D-V_G curves were measured

Table 2 Gate parameters of the analyzed Si:HfO$_2$-based MFIS-FeFET devices

	Gate length (L_G)	Gate width (W_G)
Device 1	500 nm	2 μm
Device 2	260 nm	2 μm
Device 3	100 nm	2 μm
Device 4	32 nm	1 μm
Device 5	32 nm	80 nm

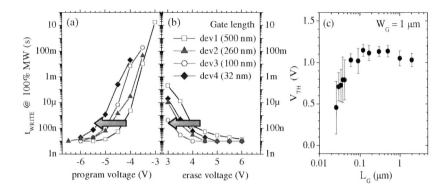

Figure 7.17 Impact of gate length scaling on the switching speed. Pulse width required to achieve the full memory window (t_{WRITE}) dependent on the pulse amplitude for (a) program and (b) erase operation. (c) An average V_{TH} value as a function of the gate length (L_G) for a constant gate width (W_G) of 1 µm.

for the unstressed single transistors with gate lengths varying from 2 µm down to 30 nm and a constant gate width of 1 µm. The reduction in the gate length induced a drop in the V_{TH} value by 0.5 V (Figure 7.17 (c)), which indicated the threshold voltage roll-off effect [259]. With decreasing gate length the effect of source and drain depletion regions on the channel charge increases, which results in lower V_{TH}-value. The V_{TH} roll off is commonly minimised by adjustment of the channel implant profiles, which enables the formation of shallow source/drain junction extensions [260], [261]. This step was left out during fabrication of the studied devices, which led to a strong source/ drain extension diffusion and the observed V_{TH} roll-off. This effect explains the detected shift in the operation voltages with decreasing gate length (Figure 7.17). The switching speed of the ferroelectric polarization is determined by the electric field in the ferroelectric layer. A shift of the intrinsic V_{TH} to more negative voltages causes a field reduction in the gate stack at negative voltages (program operation) and field enhancement at positive voltages (erase operation). Therefore, shorter channel devices could be erased at lower voltages, but required higher program voltages. The V_{TH} roll-off and, thus, the difference in the switching characteristics between long and short channel devices can be eliminated by carefully adjusting of the channel implants profiles [260], [261]. For the 32 nm FeFETs the erase characteristic did not shift to the lower voltages as expected. Scaling below 100 nm gate length seems to have some additional effects on the device performance, which is not clearly understood so far. Nevertheless, even for the 32 nm devices the erase speed in the nanosecond range can be achieved using $4-5$ V.

The endurance properties of the large (device 2: L_G = 260 nm, W_G = 2 µm) and small (device 5: L_G = 32 nm, W_G = 80 nm) devices were compared (Figure 7.18). The measurements were performed using the program/erase with width of 100 ns and amplitudes of – 6 V and + 4 V, respectively. Both devices demonstrated a comparable overall endurance of 10^4 cycles with residual MW values of 0.9 V for the 260 nm devices and 0.5 V for the 32 nm devices. In both cases the same trend in the evolution of V_{TH} values with cycling was observed. A reduction of the MW with cycling was mainly determined by the positive V_{TH} shift in the "ON" state, the V_{TH} of the "OFF" state, on the other hand, remained almost unaffected.

The data retention behaviour was evaluated at room temperature for the same devices (device 2 and 5). The experimental data were collected up to 10 days. The erased ("ON") and programmed ("OFF") states were set using +4 V/ 100 ns and – 6 V/100 ns pulses, respectively. Figure 7.19 gives the comparison between the retention behaviour of large (260 nm) and small (32 nm) FeFETs for both memory states as well as the time dependence of their MW. The evolution of the cell's V_{TH} in the "ON"/ "OFF" memory states with time changed with cell size scaling. For the 260 nm devices (device 2) more rapid V_{TH} shift was observed in the "OFF" state, while the "ON" state remained rather stable. A reverse behaviour was characteristic for the 32 nm devices (device 5) with the more stable "OFF" state and degrading "ON" state. The depolarisation field (E_{DEP}), appearing due to insufficient screening of the polarisation charge at the substrate site, is one of the major causes of the retention decay in ferroelectric transistors [110], [248]. The E_{DEP} value and, thus, retention properties depend on the intrinsic V_{TH} of the transistor [262], which is determined by the substrate doping and the work function difference. In the 32 nm devices studied in this work both memory states were shifted to more negative voltages in comparison to the 260 nm devices due to the V_{TH} roll-off effect. This shift leads to a decrease/increase of E_{DEP} in the "ON"/ "OFF" state and, hence, must have resulted in the improved/deteriorated retention for the corresponding memory states [262]. Behaviour contrary to this assumption was, however, observed (Figure 7.19). Therefore, an additional mechanism besides E_{DEP} may have determined the retention of the studied Si:HfO$_2$-based FeFETs. Charge trapping from the transistor channel is another mechanism impairing retention [110], [92], which can become crucial in the studied devices due to an interfacial layer of only 1.2 nm and high intrinsic trap density of HfO$_2$. Electron injection is to be expected during "ON" state retention. The p-substrate surface was in inversion during waiting time at zero volts due to a negative V_{TH} in the "ON" state. Electrons trapped within the HfO$_2$ compensated the positive polarisation charge, which resulted in a positive V_{TH} shift during retention. Lower "ON" state V_{TH} corresponded to higher electron density at the surface at zero volts, which enhanced trapping and degraded retention. This explains the worsening of the "ON" state retention for 32 nm devices. During the "OFF" state retention hole trapping becomes more critical. In this case, the p-substrate surface is in accumulation at zero volts. Increase in the positive V_{TH} enhances hole trapping and impairs "OFF" state retention. This behaviour was observed for 260 nm devices. The V_{TH} value of a transistor can be tuned by adjusting the channel implant profiles. In this way a trade-off between the stability of the two memory states can be found and eliminate the difference in the retention behaviour of both device types. Despite this difference for both devices a non-volatility of data up to 10 years can be projected based on the extrapolation of the experimental data. Comparable values of the residual *MW* of about 0.8 V after 10 years potential data storage were extracted for 260 nm and 32 nm devices.

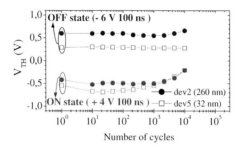

Figure 7.18 Impact of scaling on the endurance properties of Si:HfO₂-based MFIS-FETs. Endurance characteristics for two device types: device 2 (L_G = 260 nm, W_G = 2 μm) and device 5 (L_G = 32 nm, W_G = 80 nm). Similar program (– 6 V/ 100 ns) and erase (+ 4 V/ 100 ns) conditions were used for both device types.

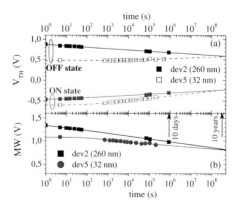

Figure 7.19 Impact of scaling on the retention behaviour of Si:HfO₂-based MFIS-FETs. (a) Retention behaviour of the "ON" and "OFF" memory states and (b) time dependence of the memory window for two device types: device 2 (L_G = 260 nm, W_G = 2 μm) and device 5 (L_G = 32 nm, W_G = 80 nm).

It should be noted that main emphasis of the present work was to examine the principle functionality of the scaled devices. Therefore, all electrical measurements, shown in this chapter, were performed on structures containing multiple ferroelectric transistors connected in parallel. In this way an averaged response from the multiple memory cells was obtained. Another important aspect of scaling, which is not considered here, is its impact on the uniformity of properties of single devices. This is especially crucial in devices including polycrystalline materials [263], which was the case for the studied Si:HfO$_2$ ferroelectric films (Figure 5.2 (a)).

7.6 Summary

The electrical behaviour of Si:HfO$_2$-based MFIS-FET devices, which were fabricated using the state-of-the-art 28 nm high-k metal gate CMOS technology, was studied in detail. The key memory characteristics such as the program and erase behaviour, retention and endurance were analysed.

The Si:HfO$_2$ film composition was shown to essentially affect the device performance (chapter 7.1). The observed trend correlated well with the results obtained on Si:HfO$_2$-based, MFM capacitors (chapter 5.1.1). Therefore, the formation of the ferroelectric phase in Si:HfO$_2$ films, which were embedded into the studied capacitors and transistors, proceeded in the same way in spite of difference in their stack structure. A dominating ferroelectric switching was detected in FeFET devices with Si:HfO$_2$ layers containing 3.7 and 4.4 cat % Si. Devices that included HfO$_2$ films with higher Si contents showed, on the other hand, a prevailing charge trapping characteristics.

The FeFETs including Si:HfO$_2$ ferroelectric films demonstrated a program and erase capability in the nanosecond time regime (10 – 100 ns) with operation voltages of 4 – 6 V (Figure 7.4). These characteristics were superior to those of the state-of-the-art FeFET cells based on perovskite-type ferroelectric SBT films [112]. These operated at comparable voltages (4 – 7 V), required, however, significantly longer program/erase times over several hundred milliseconds to achieve a memory window demonstrated for the devices studied in this work.

The operation capability of the Si:HfO$_2$-based memory cells in the temperature range between 25 and 210 °C was proven (chapter 7.3). Two distinguishable memory states and a residual memory window obtained by extrapolation to 10 years could be demonstrated at all operation temperatures (Figure 7.5). An increase of temperature deteriorated the retention properties, causing an acceleration of the V_{TH} shift with time and, as a result, a decrease in the residual MW predicted for 10 years storage. Higher operation voltages, on the other hand, improved the retention behaviour (Figure 7.6), which was, however, at the expense of the memory window size.

Endurance capability of the studied devices was worse than it could be expected from the results of the MFM capacitors testing (chapter 6.3) and was limited to 10^4 – 10^5 program/erase cycles. A correlation between the reduction of the memory window and increase of the gate leakage current was found with increasing number of endurance cycles (Figure 7.8). A degradation of the dielectric stack in MFIS-FET devices differed, however, from that of the MFM structures. No evidence of the increased trap density in Si:HfO$_2$ films was found (Figure 7.11) in contrast to capacitor structures (chapter 6.3). The interfacial SiON layer, on the other hand, degraded upon cycling and was likely mainly responsible for the deterioration of the gate leakage current. A strong increase in the trap density of the interfacial layer was confirmed by the charge pumping measurements (Figure 7.10). This predominant degradation of the interfacial layer of the gate stack, which was observed for the studied ferroelectric cells, was similar to the behaviour reported for the standard high-*k* metal gate stacks [150], [190], [252], [253], [254]. Based on the similarity of the gate stack structure of these devices a similarity also in their degradation mechanism can be assumed. In a standard high-*k* metal gate stack the wear-out of the interfacial layer is argued to be driven by a continuous charge transport [253], [254]. This charge transport was shown to be further enhanced in a gate stack that includes a ferroelectric high-*k* layer due to the presence of the ferroelectric polarisation charge (Figure 7.14). A deterioration of the insulating properties of the interfacial SiON layer may be responsible for the endurance degradation of the Si:HfO$_2$-based MFIS-FET devices as proposed in [250]. A build-in of negative charges was suggested in this work as another possible cause of the observed endurance behaviour. It can be a result of the generation of negatively charged defects within the interfacial layer or accumulation of electron trapped on the pre-existing traps of the interfacial SiON or Si:HfO$_2$ layers. Although the endurance characteristics of the studied Si:HfO$_2$-based FeFETs were inferior to those of devices with perovskite ferroelectric materials , which can withstand up to 10^{12} cycles [105], they still are able to meet the modern requirements of the Flash memories (10^4 – 10^5 program/erase cycles [4]).

The impact of scaling of Si:HfO$_2$-based MFIS-FETs down to the gate length of 28 nm on their memory performance investigated. The scaled devices demonstrated characteristics comparable to the long channel structures: program and erase times in the range of several nanoseconds (down to 10 ns) with voltages of 4 – 6 V, endurance capability up to 10^4 cycles and a comparable residual *MW* of 0.8 V projected after 10 years at room temperature. The detected differences in the behaviour between the long and short channel devices, such as shift of operation voltages and altered retention behaviour, could be, for the most part, attributed to transistor short channel effects (here V_{TH} roll-off). Therefore, a careful adjustment of the channel implant profiles in scaled cells is expected to provide behaviour similar to long-channel devices.

In addition, it was shown that charge trapping had a strong influence on the performance of the studied FeFETs. Charge trapping was detected to superimpose with ferroelectric switching during program and erase operations, which resulted in degradation of the ferroelectric memory window (chapter 7.2). Time dependent V_{TH} shift observed during retention tests for both long and short channel devices could be explained in terms of the charge trapping (chapter 7.5). Moreover, charge trapping may be the main cause of the endurance degradation in Si:HfO$_2$-based MFIS-FET structures (chapter 7.4). Since trapping was shown to affect all the key characteristics of the Si:HfO$_2$-based memory cells, the traps in the MFIS gate stack and associated trapping phenomenon were studied in more details. The results of these analyses are presented in the next section.

8   Trapping effects in Si:HfO$_2$-based FeFETs

HfO$_2$-based materials are known to exhibit a high intrinsic defect densities (10^{12} – 10^{14} cm^{-2}) [141], [142], which act as electron and/or hole traps [141], [144], [145], [146]. A review of HfO$_2$ traps and their characteristics was given in chapter 2.3.4. For the same polarity of the gate voltage the charge trapping induces a V_{TH} shift, which is reverse to that of caused by the ferroelectric switching. Therefore, the charge trapping in a FeFET memory cell has a negative effect on its performance. Charge trapping was one of the issues in the first perovskite-based FeFET devices [101], [102], in which the ferroelectric layer was fabricated directly on the semiconductor substrate. The V_{TH} shift observed in these devices was dominated by the charge trapping that masked the ferroelectric polarisation. Thick interfacial layers of 5 – 10 nm are used in the contemporary FeFETs with perovskite ferroelectrics [111], [112], which reduce the trapping, however, at the expense of the gate scaling capability.

The traps in the gate stack of the Si:HfO$_2$-based MFIS-FET cells and their impact on the memory performance of these devices were analysed in this work. The trapping (chapter 8.1) and detrapping (chapter 8.2) characteristics were investigated by means of a single-pulse I_D-V_G technique (chapter 3.3.2). The impact of trapping on the performance of the studied Si:HfO$_2$-based FeFET cells is discussed in chapter 8.3. A modified approach for erase operation, which should mitigate the impact of trapping is proposed in chapter 8.4.

8.1 Trapping kinetics of the bulk Si:HfO₂ traps

The trapping behaviour of Si:HfO$_2$-based MFIS-FET structures was investigated by means of the single-pulse measurement technique (chapter 3.3.2). The stress pulse of + 4 V, which corresponded to a typical erase voltage required to set the "ON" memory state (chapter 7.2), was applied to the gate. The drain current was simultaneously monitored and translated into the I_D-V_G characteristics (Figure 8.1 (a)). The pulse width was varied was varied between 0 and 50 μs, whereas the rise and fall times were constant and amounted to 500 ns. All pulses started at – 4 V to ensure a complete discharge of traps. The drain voltage was set to 300 mV. This enabled to maximise the signal-to-noise ratio due to increased drain current. A positive shift between the I_D-V_G curves measured at the rising and falling edges was observed (Figure 8.1 (a)). This shift as well as a decrease of the drain current during the pulse maximum (Figure 8.1 (b)) indicated a strong electron trapping. A flattening of the I_D-V_G characteristic was observed at the rising pulse edge for gate voltages higher than 2 V (Figure 8.1 (a)), which can also be attributed to the fast electron trapping. The shift between the I_D-V_G curves of the rising and falling edge (ΔV_{TH}) is proportional to the amount of trapped electrons. ΔV_{TH} and, thus, an amount of trapped electrons increased for longer stress pulses (inset Figure 8.1 (b)). A logarithmic time dependence of ΔV_{TH} indicated that the tunneling processes was the dominating injection mechanism [142], [264]. A positive V_{TH} shift, which induced by positive

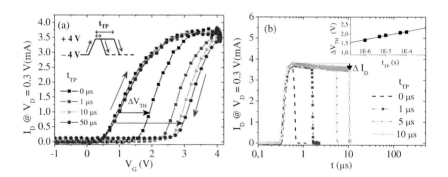

Figure 8.1 Single-pulse characteristics of the Si:HfO$_2$-based MFIS-FET structures: (a) I_D-V_G characteristics measured on the rising and falling edges of the excitation gate pulse for varying pulse width. The experimental excitation gate pulse is shown in the inset. (b) Evolution of I_D during the excitation gate pulse for varying pulse width. Inset: V_{TH} shift as a function of the gate pulse width.

gate pulses, observed using the single-pulse technique (Figure 8.1 (a)) was in contradiction to the results of the chapter 7.2. A negative V_{TH} shift was shown for positive erase pulses, whereas a positive V_{TH} shift was induced by negative gate pulses (Figure 7.3 (a)). This contradiction can be explained by taking into account the difference between the measurement procedures, which were used during both tests. In the single-pulse measurement the I_D-V_G characteristics were recorded directly at the rising and falling edges of the stress pulse. In this way the time delay between stressing and sensing was practically eliminated and fast trapping effects could be observed. During testing of the erase operation of the FeFET cells (chapter 7.2), on the other hand, the I_D-V_G characteristics were measured with a time delay of several seconds after the positive erase pulse, when the detrapping processes have set in so that a negative V_{TH} shift was observed.

In order to identify the operation conditions (pulse width and amplitude), which enable to eliminate the electron trapping for positive pulses, single-pulse measurements were performed using the gate pulse sequence illustrated in Figure 8.2 (b). Two identical positive single-pulses were applied consecutively to the gate with a time delay of 1 min. This time delay was required for the detrapping processes to set in. A negative polarisation state was re-established each time before single-pulses by applying an initialisation pulse of – 6 V for 100 ns. The results of the measurement for the single-pulses with varying width and amplitude are depicted in Figure 8.2 (b). Here, the V_{TH} shifts at the falling edge of the first single-pulse ($\Delta V_{TH}12$) and at the rising edge of the second single-pulse ($\Delta V_{TH}13$) are plotted versus the pulse width for different pulse voltages. For all tested erase pulse conditions the $\Delta V_{TH}12$ was positive. Thus, the ferroelectric switching was accompanied by the electron trapping for all erase conditions. This trapped electron charge exceeded the ferroelectric polarisation charge directly after the pulse. The ferroelectric switching was detected for pulse voltages above 3 V, for which negative $\Delta V_{TH}13$ after a delay of 1 min was observed. Therefore no erase operation conditions could be found, for which ferroelectric switching occurred without simultaneous electron trapping. Another interesting fact was detected by comparing the $\Delta V_{TH}12(t_{TP})$ and $\Delta V_{TH}13(t_{TP})$ characteristics for 3 V. The start of a strong electron trapping ($\Delta V_{TH}12$ became positive) corresponded with the completion of the ferroelectric switching ($\Delta V_{TH}13$ saturated). In order to clarify this phenomenon the energy band diagram of the studied gate stack was calculated for 4 V gate voltage in case of a negatively ($P_R = - 2$ µC/cm^2) and positively ($P_R = + 2$ µC/cm^2) polarised Si:HfO$_2$ layer (Figure 8.2 (c)). The multi-dielectric energy band diagram program [257], [258] was used for this purpose. It can be seen that the polarisation charge of the Si:HfO$_2$ layer gives rise to the internal electric fields and affects the band bending at the SiON-Si:HfO$_2$ interface. This alters the effective thickness of the tunneling barrier and, thus, the injection probability.

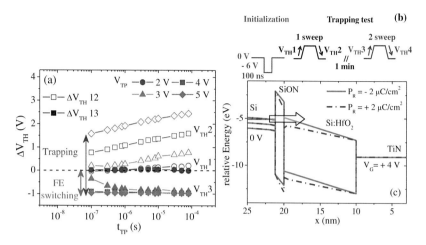

Figure 8.2 Superposition of ferroelectric switching and trapping for positive gate pulses studied using the single-pulse technique: (a) V_{TH} shift directly after the pulse ($\Delta V_{TH}12$) and after 1 min delay ($\Delta V_{TH}13$) as a function of the pulse width for different pulse voltages; (b) Experimental gate pulse sequence; (c) Energy band diagram of a Si/SiON/Si:HfO$_2$/TiN stack in case of a positive ($P_R = + 2\ \mu C/cm^2$) and a negative ($P_R = - 2\ \mu C/cm^2$) ferroelectric polarisation charge. The ferroelectric polarisation charge affects the effective thickness of the tunneling barrier and, thus, the trapping probability.

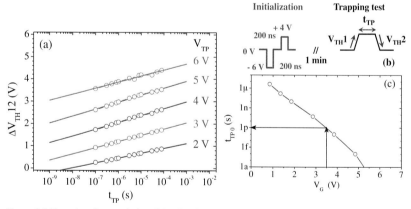

Figure 8.3 Trapping characteristics of the Si:HfO$_2$ MFIS-FET structures in appositively polarised state used to estimate the trapping onset time: (a) ΔV_{TH} between rising and falling pulse edges as a function of the pulse width for varying gate voltages: experimental data (symbols) and fit (lines); (b) Experimental gate pulse sequence; (c) The onset time (t_{TP0}) as a function of the gate pulse voltage.

In case of a positive polarisation charge the bands are lowered, which decreases the effective thickness of the tunneling barrier. This enhances the injection probability, which, in turn, results in the increased amount of the trapped charge. A negative polarisation charge has a reverse effect on the injection probability. The bands are shifted upwards, so that the effective thickness of the tunnelling barrier increases and the injection is impeded. This explains the experimental results, showing a strong electron trapping as soon as the Si:HfO$_2$ was switched into a positive polarisation state (Figure 8.2 (b)). This also means that a positive polarisation charge established during the positive erase pulses enhances the deleterious electron trapping.

In order to estimate the pulse width, which corresponds to the onset of the trapping process, the trapping behaviour of the MFIS-FET structure with a positively polarised Si:HfO$_2$ layer was analysed. The single-pulse measurements were performed using the pulse sequence shown in Figure 8.3 (b). Before each trapping pulse a positive polarisation state was re-established with a combination of a negative ($-$ 6 V for 200 ns) and a positive ($+$ 4 V for 200 ns) pulses. A time delay of 1 min between the initialisation pulse sequence and a trapping pulse should enable the detrapping processes to set in. The resulting trapping characteristics, V_{TH} difference between the rising and falling pulse edges, are shown as a function of the trapping pulse width for varying pulse voltages (Figure 8.3 (a)). The experimental data (symbols) were extrapolated (lines) to the zero $\Delta V_{TH}/2$. The intersection with the time axis gives the pulse width corresponding to the onset of trapping ($t_{TP\,0}$). This $t_{TP\,0}$ is equivalent to the critical trapping time introduced in [149]. An increase in the gate voltage causes the trapping start at shorter pulse widths (Figure 8.3 (c)). In case of gate voltages between $+$ 3.5 V and $+$ 4 V, which were identified as optimal erase conditions for the studied FeFET memory cells (chapter 7.2), pulses shorter than several picoseconds will be required to eliminate any trapping.

8.2 Detrapping kinetics of the bulk Si:HfO$_2$ traps

It has been shown in the previous chapter that the separation of the ferroelectric switching and trapping during erase operation is impossible to realise. Therefore, the detrapping of electrons, which were trapped during a positive erase pulse, was studied in order to estimate how fast the detrapping processes set in and the ferroelectric memory window can be sensed. The characteristics detrapping times for gate stacks comparable to the FeFET devices, which are studied in this work, were shown to be in the microsecond time range [264], [265], [266], [267], [196]. Therefore, a single-pulse technique was chosen again as the most suitable method for the characterisation of detrapping in Si:HfO$_2$-based MFIS-FET devices after positive erase pulses. The pulse sequence used for the discharging analyses is illustrated in Figure 8.4 (b). Two identical positive single-pulses with a voltage

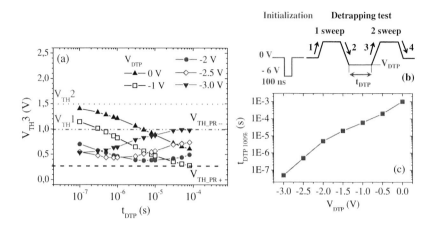

Figure 8.4 Detrapping characteristics for Si:HfO$_2$-based MFIS-FET devices after a positive erase pulse of +3.5 V for 300 ns: (a) $V_{TH}3$ after a detrapping delay as a function of the detrapping time (t_{DTP}) for varying detrapping voltages (V_{DTP}); (b) Experimental gate pulse sequence; (c) Voltage dependence of the detrapping time, which is required to achieve a complete detrapping ($t_{DTP\ 100\%}$).

of + 3.5 V and width of 300 ns were applied consecutively to the gate. A time delay between these pulses (t_{DTP}) as well as a voltage applied during this delay (V_{DTP}) was varied and the induced V_{TH} shift was characterised. Initialisation pulse of – 6 V was applied for 100 ns each time before the single-pulses. It ensured the detrapping of residual trapped electrons on the one hand and, on the other, set an identical initial cell state with a negative polarisation ($V_{TH}1$). The voltage and width of the single-pulses was chosen so that they were close to parameters of the erase pulses used for Si:HfO$_2$-based FeFET memory cells (chapter 7.2). The results of the detrapping measurements are presented in Figure 8.4 (a). The value of $V_{TH}3$, which was obtained at the rising edge of the second single-pulse, is plotted as a function of the detrapping time (t_{DTP}) for different detrapping voltages (V_{DTP}). The threshold voltage of the studied devices in the negatively polarised state (V_{TH_PR-}) was defined as $V_{TH}1$ at the rising edge of the first single-pulse. The threshold voltage in the positively polarised state (V_{TH_PR+}) was determined from previous measurements (Figure 8.2 (a)). $V_{TH}2$, which was measured directly after the first single-pulse at its falling edge, was shifted to more positive voltages in respect to the $V_{TH}1$. This positive shift indicated electron trapping into the gate stack superimposed with the ferroelectric switching. After the detrapping period a change in the direct of the V_{TH} shift was observed. A continuous decrease of $V_{TH}3$ with increasing time delay between the first and the second single-pulse can be attributed to the detrapping of previously trapped electrons. The back tunneling of electrons from the Si:HfO$_2$ traps into the Si conduction band is the most probable discharging mechanism as suggested for comparable gate stacks in [264]. The observed logarithmic time dependence of $V_{TH}3$ (for V_{DTP} of 0 and − 1 V) can be explained by the spatial distribution of traps within the Si:HfO$_2$ layer. Traps located closer to the SiON-Si:HfO$_2$ interface exhibit shorter characteristics detrapping times than traps located deeper in the high-*k* layer. With progressive detrapping $V_{TH}3$ approached the value corresponding to the positive polarisation state (V_{TH_PR+}). The time of the complete detrapping ($t_{DTP\ 100\%}$) was assigned to the time, when $V_{TH}3$ becomes equal to V_{TH_PR+}. A clear enhancement of the detrapping rate was detected for more negative V_{DTP} (Figure 8.4 (c)). At a voltage of 0 V a complete detrapping can be achieved only after a few milliseconds. By applying voltages above − 2.5 V and − 3 V this time can be reduced to a few hundreds of nanoseconds. These voltages can be, however, critical if they applied for a longer time. An increase of $V_{TH}3$ was detected for these voltages after some time, which indicated a reversal of the ferroelectric polarisation from the positive into negative.

8.3 Impact of trapping on the FeFET performance

Strong trapping under typical operation conditions was demonstrated for Si:HfO$_2$-based MFIS-FET devices in the chapter 8.1. This chapter will be devoted to a detailed discussion about the impact of trapping on the performance of Si:HfO$_2$-based FeFET memory cells, in particular erase operation and endurance characteristics.

8.3.1 Erase operation

One of the advantages of the ferroelectric memories over the state-of-the-art FG memories is a significantly faster writing capability. The polarisation state of a ferroelectric device can be switched within a few nanoseconds, while charge the trapping and especially detrapping in the FG cells requires from tens of microseconds to several milliseconds [4]. The polarisation switching in the nanosecond time range was also demonstrated for Si:HfO$_2$-based FeFETs (Figure 7.4 (b)). Therefore, these devices can be theoretically operated in a GHz frequency range. Additional electron trapping during the erase pulses, which was shown to be inevitable for all operation conditions (chapter 8.1), reduces, however, considerably the effective erase speed. The cell readout cannot be performed while traps are discharged. Otherwise an erroneous data interpretation will be a result, since the trapped electrons alter the V_{TH} value of a memory cell. Thus, the effective erase time (t_{ERASE_eff}) should include besides the time required for the polarisation switching (t_{SWITCH}) also the time of electron detrapping (t_{DTP}). At zero volts a complete detrapping could be achieved only after several milliseconds (Figure 8.4 (c)). Therefore, the t_{ERASE_eff} will also lie in the millisecond range.

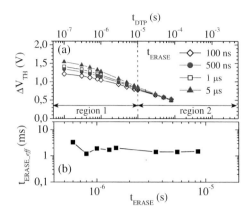

Figure 8.5 Impact of the erase pulse width on the subsequent electron discharging. (a) Time dependent V_{TH} shift at V_{DTP} of 0 V, caused by electron detrapping, after an erase pulse of +3.5 V with varying width; (b) Effective erase time (t_{ERASE_eff}) as a function of the erase pulse width.

The amount of trapped electrons can be reduced by using shorter erase pulses. The impact of erase pulse width on the subsequent detrapping behaviour and on the effective erase time was analysed (Figure 8.5). Time dependent V_{TH} shift, caused by electron detrapping, was monitored at V_{DTP} of 0 V after an erase pulse of +3.5 V with width varying from 100 ns to 5 μs (Figure 8.5 (a)). The erase pulse width was varied between 100 ns and 5 μs. Shorter erase pulses could not be tested due to the limitation of the experimental measurement setup. The longer erase pulses resulted in higher V_{TH} values due to more trapped electrons. In the region 1 with detrapping times below 10 μs the detrapping rate showed a clear dependence on the erase pulse width. The detrapping was faster for longer preceding erase pulses. This can be explained by higher built-in electric fields, which arose from higher amount of trapped charges. This higher build-in electric field caused enhance detrapping. In the region 2, for detrapping times above 10 μs, however, the detrapping characteristics for all erase pulses merged together and revealed a similar detrapping behaviour. As a result, the time required for detrapping and, thus, the effective erase time turned out to be independent of the erase pulse width (Figure 8.5 (c)). Therefore, the trapping can be tolerated only to the amount that still enables to distinguish the erased ("ON") from the programmed ("OFF") cell state. In this case the detrapping step will be unnecessary, so that the effective erase time will include only the time of the ferroelectric switching ($t_{ERASE\_eff} = t_{SWITCH}$). If the amount of trapped electrons, however, exceeds the above mentioned condition and two "ON" and "OFF" state cannot be distinguishable, the effective erase times of several milliseconds should be expected independent on the absolute amount of trapped charges. The erase pulse width ($t_{TP\,0}$), required to eliminate the trapping in the studied Si:HfO$_2$ FeFET devices, were estimated in chapter 8.1. For typical erase voltages between + 3.5 V and + 5 V, pulses shorter than several picoseconds will be required to eliminate any trapping ((Figure 8.3 (c))). A possible solution to this trapping issue will be discussed in the chapter 8.4, where a modified erase pulse is proposed. It consists of a pulse responsible for the reversal of the ferroelectric polarisation in combination with a pulse, which accelerates the detrapping process.

8.3.2   Endurance behaviour

Endurance characteristics of Si:HfO$_2$-based FeFET devices were discussed previously in chapter 7.4. Both scenarios suggested as explanation for the endurance degradation in the studied devices were based on the assumption of electron trapping, which accompanies the ferroelectric switching. An experimental evidence of a strong electron trapping at typical erase operation conditions was demonstrated in chapter 8.1. This justifies the proposed endurance models. Fast trapping rates at positive gate voltages (Figure 8.3) as well as detrapping rates at negative gate voltages (Figure 8.4) comparable with the rates of the ferroelectric switching were ascertained. This confirm an assumption of a continuous charge transport through the interfacial SiON layer during program/erase cycling, which explains the observed wear-out of the interfacial layer and degradation of the gate leakage.

8.4   Modified approach for erase operation

A fast erase operation of the Si:HfO$_2$-based FeFET memories is impeded by the electron trapping and their slow detrapping at zero volts as discussed in the chapter 8.3.1. Fast trapping rates competing with the rates of the ferroelectric switching make the complete elimination of trapping for investigated devices impossible. In order to enable a fast erase operation, one can try to accelerate the detrapping of captured electrons. This can be done by applying a negative gate voltage (Figure 8.4). Taking this into account a modified erase pulse is proposed (Figure 8.6).

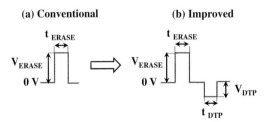

Figure 8.6 Conventional (a) and modified (b) form of an erase pulse for FeFET memory cells.

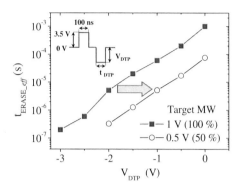

Figure 8.7 Impact of the detrapping pulse voltage (V_{DTP}) on the effective erase time (t_{ERASE_eff}) for different target values of *MW* directly after the erase pulse: 1 V corresponding to the maximum achievable *MW* value and 0.5 V corresponding to a half of the maximum achievable *MW* value. Inset shows the erase pulse used.

In contrast to a conventional erase operation a modified erase pulse includes a positive pulse, which sets a positive polarisation state in the ferroelectric, and a negative pulse, which accelerates the detrapping of captured electrons. This modified erase pulse enables to sense the true V_{TH} state of a FeFET cell directly after the erase operation without any time delays required. In this case the effective erase time (t_{ERASE_eff}) is the sum of switching (t_{ERASE}) and detrapping (t_{DTP}) pulse width. Figure 8.7 illustrates the impact of the detrapping pulse voltage (V_{DTP}) on the effective erase time. The first switching pulse exhibited a constant voltage of 3.5 V and width of 100 ns. The width of the second detrapping pulse was adjusted for each detrapping voltage so that either the maximum *MW* of 1 V or a half of the maximum *MW* (0.5 V) was achieved directly after the erase operation. The results of the detrapping studies presented in chapter 8.2 were used for this purpose. The effective erase time decreased for higher voltages of the detrapping pulse (Figure 8.7). With detrapping voltages between – 2.5 and – 3 V t_{ERASE_eff} can be reduced to several hundred nanoseconds. If a half-value of the maximal *MW* (0.5 V) instead of the maximum *MW* of 1 V is targeted after the erase pulse the same t_{ERASE_eff} can be provided at lower detrapping voltages. Thus, the erase operation of the studied FeFET devices in the nanoseconds time can be enabled by using the modified erase pulses. The shortest t_{ERASE_eff} demonstrated in this work was 200 ns. This is due to the limitations of the experimental measurement setup, which limits the width of the pulses to 100 ns. Even better results with shorter t_{ERASE_eff} are expected for the measurement setup allowing application of shorter pulse width.

In addition, the impact of the modified erase pulse form on the endurance behaviour of the Si:HfO$_2$-based FeFETs was investigated (Figure 8.8 and Figure 8.9) in order to gain better insight into the correlation between the electron trapping and endurance degradation. In the first step the impact of discharging at 0 V on the endurance behaviour was analysed (Figure 8.8). All endurance measurements were performed using alternating erase and program pulses of +3.5 V and – 6 V, respectively, applied for 300 ns. The detrapping step at 0 V of a varying duration (t_{DTP}) from 100 ns to 1 s was included directly after the positive erase pulse (Figure 8.8 (a)). All devices were pre-cycled with the same conditions before the endurance test in order to obtain comparable values of the memory window. The resulting endurance characteristics are shown in Figure 8.8 (b). The degradation of the V_{TH} in the "ON" state slowed down for stress conditions with longer detrapping steps, which corresponded to higher degree of detrapping. The measurements with the longest t_{DTP} of 1 s even showed an increase of the *MW* up to 10^3 program/erase cycles. With further cycling, however, even more rapid degradation than for the other cases was observed. Furthermore, the endurance behaviour with the erase pulses in a modified form (Figure 8.6 (b)) was characterised. A negative detrapping pulse was included after the positive erase pulse (Figure 8.9 (a)).

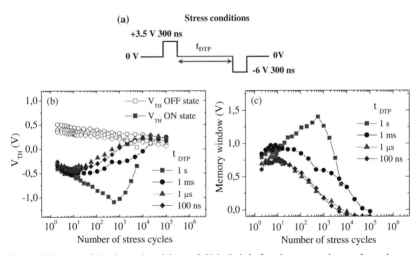

Figure 8.8 Impact of the detrapping delay at 0 V included after the erase pulse on the endurance characteristics of the Si:HfO$_2$-based FeFET devices. (a) Experimental gate pulse sequence used for the endurance testing including a detrapping step at 0 V. Evolution of (b) the V_{TH} in the "ON" and "OFF" memory states and (c) a resulting memory window with increasing number of stress cycles for test pulses with varying detrapping times.

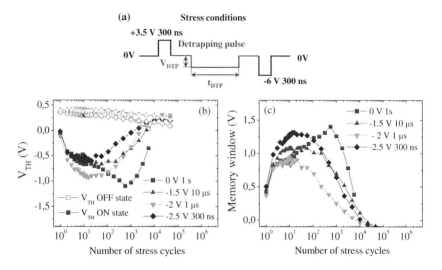

Figure 8.9 Impact of the detrapping step at negative voltages included after the erase pulse on the endurance characteristics of the Si:HfO$_2$-based FeFET devices. (a) Experimental gate pulse sequence used for the endurance testing including a negative detrapping pulse. Evolution of (b) the V_{TH} in the "ON" and "OFF" memory states and (c) a resulting memory window with increasing number of stress cycles for test pulses with different detrapping pulses.

Detrapping pulses with different voltages were tested. The width of the detrapping pulse was chosen dependent on its voltage so that it was equal to the corresponding $t_{DTP\ 100\%}$ value (Figure 8.4 (c)). Although for all test conditions a complete detrapping of electrons captured during the positive pulse was provided by the detrapping pulse, a difference in the endurance behaviour for different stress conditions was observed. Higher negative detrapping voltages aggravated the endurance degradation of the studied devices. Thus, it can be deduced, that the degradation mechanism of endurance in the Si:HfO$_2$-based FeFET devices must be more complicated than just an accumulation of trapped electrons within the Si:HfO$_2$ film.

8.5 Summary

Application of a single-pulse measurement technique has enabled to obtain an experimental evidence of a strong electron trapping in Si:HfO$_2$-based MFIS-FET devices during the erase operation (chapter 8). At typical erase operation conditions (i.e., pulses of 3.5 – 5 V with a width of 100 ns to 10 µs) the electron trapping was shown to accompany the ferroelectric switching (Figure 8.2 (a)). Moreover, directly after the erase pulse the ferroelectric memory window was masked by the charge of the trapped electrons, which exceeded the ferroelectric polarisation charge. The reason for such this severe trapping is a thin interfacial SiON layer of the studied gate stack (1.2 nm) but also the polarisation charge of the Si:HfO$_2$ layer. The latter induces internal electric fields, which enhance the electron injection (Figure 8.2 (c)). A correlation between the start of the electron trapping and ferroelectric switching was found. The estimated characteristic trapping times lay in the range of sub-nanosecond (Figure 8.3 (c)). Therefore, a complete elimination of trapping during erase operation is impossible for the studied gate stacks.

Detrapping characteristics of electrons captured in Si:HfO$_2$ layer during the erase pulse have been investigated (chapter 8.2). At zero volts the detrapping times of several milliseconds are required to sense the ferroelectric memory window (Figure 8.4 (c)). This slow detrapping of the captured electrons impedes a fast erase of the Si:HfO$_2$-based FeFET memories (chapter 8.3.1). Besides the time required for the polarisation switching the time required for the detrapping processes to set in determines the effective erase time. The width of the erase pulse was shown to have a negligible influence on the overall detrapping time (Figure 8.5). Therefore, the trapping can be tolerated only to the amount that still enables to distinguish the erased ("ON") cell state from the programmed ("OFF") cell state. Otherwise the effective erase times will achieve several milliseconds independent on the absolute amount of the trapped charge.

In order to enable a fast erase operation, a modified erase pulse form has been proposed (chapter 8.4). It makes use of the fact that the electron detrapping rate can be accelerated by applying negative gate voltages (Figure 8.4). The proposed erase pulse consists of two consecutive pulses – a positive pulse, which induces a positive ferroelectric polarisation, and a negative pulse, which accelerates the detrapping of captured electrons (Figure 8.6). This modified erase operation enabled to reduce the effective erase time to nanosecond range (Figure 8.7). The impact of the electron detrapping on the endurance characteristics of the Si:HfO$_2$-based MFIS-FET devices has been analysed in order to gain more insight into the correlation between the trapping and the endurance degradation (chapter 8.4). A slowing down of the endurance degradation was observed when electrons captured during the erase pulse had sufficient time to be detrapped before the subsequent program pulse (Figure 8.8). More important than the detrapping capability is, however, the conditions under which this process occurs (Figure 8.9). A slow detrapping at low negative voltages was found to be more advantageous than a fast detrapping at high negative voltages. Therefore, the endurance degradation of the Si:HfO$_2$-based FeFET devices is not only limited by the accumulation of trapped electrons within the Si:HfO$_2$ film. The deterioration of the interfacial SiON layer, as proposed in chapter 7.4, must also have an important effect on the endurance behaviour of the studied devices.

9 Summary and Outlook

The ferroelectric field effect transistors (FeFETs) are considered as a promising candidate for future non-volatile memory applications [21], [14], [94] due to their attractive features, such as the non-volatile data storage, program/erase times in the range of nanoseconds, low operation voltages, almost unlimited endurance (above 10^{12} cycles [105]), non-destructive read-out and a compact one-transistor cell structure without any additional access device needed. Despite the efforts of many research groups an industrial implementation of the FeFET concept is still missing. The main obstacles originate from the conventional perovskite-type ferroelectric materials (lead zirconium titanate (PZT) and strontium bismuth tantalate (SBT)), in particular their integration [16], [17] and scaling issues [13]. The recently discovered ferroelectric behaviour of HfO_2-based dielectrics [18], [19], [20] yields the potential to overcome these limitations. The decisive advantages of these materials are their full compatibility with the standard CMOS process and better scaling potential. Due to a significantly higher coercive field strength and lower dielectric constant in comparison to perovskite ferroelectrics the gate stack height of HfO_2-based FeFET devices can be lowered to several nanometres. This provides gate stack aspect ratios more suitable for scaling. By utilizing $Si:HfO_2$ ferroelectric thin films, FeFETs were fabricated at a state-of-the-art CMOS technology node of 28 nm, which finally closed the scaling gap between the ferroelectric and logic transistors [24].

The ferroelectric properties of HfO_2 thin films are known only since 2010 [22]. Therefore, there are still a lot of uncertainties about the origin of this ferroelectric behaviour as well as the impact of different fabrication conditions on its emergence. Moreover, the electrical behaviour of both the HfO_2-based ferroelectric films and memory devices based on these films requires more detailed studies. The emphasis of this work lay on the ferroelectric properties of HfO_2 thin films doped with silicon ($Si:HfO_2$). The potential and possible limitations of this material system with the respect to the application in non-volatile FeFET-type memories were extensively examined.

The material aspects of the Si-doped HfO_2 thin films were studied at first in order to gain better insight into the occurrence of ferroelectricity in this system and to acquire guidelines for FeFET fabrication. The influence of the different process parameters such as

Si doping concentration, post-metallisation annealing conditions and film thickness on the stabilisation of the ferroelectric properties in Si:HfO$_2$ films has been examined. Electrical characterisation combined with structural analyses enabled the changes in the macroscopic electrical properties to be correlated to alterations in the film crystalline structure. This indicates that the origin of the ferroelectric effect in HfO$_2$ lies in its crystalline structure. The ferroelectric characteristics of the Si:HfO$_2$ films were shown to appear at the phase boundary between the monoclinic (P2$_1$/c) and tetragonal (P4$_2$/nmc) phase. The transition between these two phases was induced by the silicon doping. This behaviour is similar to that of HfO$_2$ films with other dopants, e.g., Zr [19], Y [20], Al [130], and Gd [131]. A non-centrosymmetric orthorhombic phase Pbc2$_1$, which can be stabilised at the monoclinic-to-tetragonal phase boundary as claimed in [18], was also held responsible for the ferroelectric properties of the films studied in this work. In addition to the *P-V* hysteresis loops and butterfly-shaped *C-V* curves, a strong evidence for the structural ferroelectricity in Si:HfO$_2$ films was provided by PFM measurements. A distinct piezoelectric response, which is a necessary requirement for a non-centrosymmetric ferroelectric phase [69], was demonstrated in combination with the ability to locally reverse the film polarisation in an external electric field. The temperature of the post-metallisation annealing as well as the film thickness significantly affected the ferroelectric properties of Si:HfO$_2$. In both cases it correlated with a change in the fraction of the ferroelectric phase. The post-metallisation anneal affected primarily the degree of the film crystallinity. An enhancement of the P_R-value from 5 to 24 µC/cm^2 was obtained by increasing the annealing temperature from 650 to 1000 °C. The film thickness had a reverse effect on the ferroelectric properties of Si:HfO$_2$. An impairment of the ferroelectric behaviour was a result of the film thickness increased from 9 to 50 nm. This effect could be attributed to the increased stability of the monoclinic phase, which impeded the formation of a ferroelectric orthorhombic phase, leading to a reduction of the remanent polarisation. The observed stabilisation of the monoclinic phase for thicker films resulted from the combined effect of the decreased surface energy [120], [121] and insufficient mechanical stress during crystallisation. Thicker films (27 and 50 nm) crystallised already during the deposition of the top TiN electrode due to a reduced crystallisation temperature, while the thinner films (9 nm) were crystallised after they were embedded between the top and bottom TiN layers. One of the important findings of this work is that the anneal at 1000 °C for 1 s, equivalent to dopant activation anneal used during CMOS process, was sufficient for crystallisation of the Si:HfO$_2$ films and induced a ferroelectric behaviour with P_R of 24 µC/cm^2. Therefore, Si:HfO$_2$-based ferroelectric transistors can be fabricated using state-of-the-art CMOS process without the requirement for additional annealing steps. This is a real advantage in comparison to the PZT and SBT films that require special integration schemes due to the high processing temperatures (600 – 800 °C), high pressure oxygen atmosphere during deposition and the sensitivity of the ferroelectric properties to the hydrogen used during forming gas anneals

[97], [98], [17]. Utilisation of chloride-based precursors and H_2O for the fabrication of Si:HfO$_2$ films in this work allowed higher P_R-values (18 – 24 µC/cm^2) to be achieved in comparison to films grown with metal-organic precursors and O_3 (P_R from 5 to 12 µC/cm^2), which were analysed in previous works [22], [207], [229] . The values of the coercive fields (0.7 – 1 MV/cm) and the Si doping range with the most prominent ferroelectric properties (2.5 – 4 cat% Si) were, on the other hand, similar for both deposition processes. The studied Si:HfO$_2$ films exhibited P_R-values comparable to perovskite ferroelectric materials (e.g., PZT and SBT), however, a factor of ten higher coercive field strength (E_C for PZT and SBT ~50 kV/cm). This high E_C-value enabled a sufficiently high memory window to be achieved in the subsequently fabricated Si:HfO$_2$-based FeFET devices, in which the ferroelectric film thickness was reduced to 9 nm in contrast to devices with PZT and SBT films (200 – 400 nm [112], [111]).

The potential of ferroelectric Si:HfO$_2$ films for ferroelectric memory applications was studied in a further step. The effect of field cycling, polarisation switching kinetics and fatigue properties has been analysed in detail. "Wake up" effect that is often observed in perovskite ferroelectric [232], [233] was also detected for Si:HfO$_2$ ferroelectric films, indicating a high density of defects responsible for the pinning of the domain walls [56], [209]. By applying an alternating electrical stress it was possible to open the initially pinched hysteresis loops as well as to improve retention properties of the remanent polarisation. The switching capability of Si:HfO$_2$ ferroelectric films was found to be comparable to the perovskite-type ferroelectric thin films [65], [66], [67]. Switching times in the nanosecond range at voltages as low as 2 V to 4 V could be demonstrated by means of the pulsed measurement technique. Therefore, the Si:HfO$_2$-based memories can provide a significant advantage in terms of operation voltage and programming speed in comparison to the state-of-the-art floating-gate technology. The switching kinetics in Si:HfO$_2$ films were better described by the nucleation-limited-switching model [61], rather than the classical Kolmogorov-Avrami-Ishibashi switching theory [59], [60]. The polycrystalline structure of the investigated samples may be the reason for the detected switching behaviour [61], [64]. Fatigue properties of Si:HfO$_2$ films have been studied depending on frequency and voltage amplitude. At low frequencies (10 – 100 kHz) and high switching fields (above 3 MV/cm) the maximum number of switching cycles was limited to 10^4 – 10^6 by a dielectric breakdown rather than a fatigue-induced reduction of the remanent polarisation. Operation at MHz frequencies and moderate electric fields (2.5 – 3 MV/cm) enabled to extend the cycling capability to 10^9 cycles. A fatigue-free behaviour up to 10^6 cycles was demonstrated. The P_R reduction with cycling could be attributed to the defect generation. The observed fatigue properties of the Si:HfO$_2$ films were comparable to those of PZT ferroelectrics combined with Pt electrodes, in which the onset of the polarisation degradation was also reported between

$10^4 - 10^7$ switching cycles [78]. PZT films with oxide electrodes and SBT films exhibit commonly superior fatigue properties ($10^9 - 10^{12}$ cycles) [50], [240], [45], [241]. For FeFET applications the evolution of E_C with cycling is, however, of greater importance, since this is the factor, which predominantly determines the memory window [107]. Promising cycling properties with a negligible change in the coercive fields up to 10^9 cycles were obtained at moderate voltages and MHz frequencies.

The Si:HfO$_2$-based MFIS-FET devices were fabricated using the state-of-the-art 28 nm high-k metal gate CMOS technology. The key memory characteristics such as the program and erase behaviour, retention and endurance were analysed. A correlation between the performance of the FeFET structures and the capability to stabilise a ferroelectric phase in the Si:HfO$_2$ films was shown. The studied FeFETs demonstrated a program and erase times in the nanosecond time regime (10 – 100 ns) with operation voltages of 4 – 6 V. These characteristics were superior to those of the state-of-the-art FeFET cells based on ferroelectric SBT films [112]. Moreover, the operation capability of the Si:HfO$_2$-based devices was proven in the temperature range between 25 and 210 °C. Two distinguishable memory states and a residual memory window obtained by extrapolation to 10 years could be demonstrated at all operation temperatures. An increase of temperature deteriorated the retention properties, causing an acceleration of the V_{TH} shift with time and, as a result, a decrease of the residual *MW* predicted for 10 years storage. Higher operation voltages, on the other hand, improved the retention behaviour, which was, however, at the expense of the memory window size. Furthermore, the Si:HfO$_2$-based FeFETs showed a deterioration of the endurance properties in comparison to the MFM structures. The memory window closed after about $10^4 - 10^5$ program/erase cycles. The discrepancy in the cycling behaviour for the transistor and capacitor structures was attributed to a difference in the degradation of their gate stacks. No evidence of the increased trap density in the Si:HfO$_2$ films of the FeFET devices was found in contrast to capacitor structures. A predominant degradation of the interfacial SiON layer in the transistor gate stacks upon cycling was ascertained by the charge pumping measurements. This behaviour was similar to that reported for the standard high-k metal gate stacks [150], [190], [252] – [254]. A similar degradation mechanism was also assumed based on the similarity of the studied ferroelectric gate stack to the standard high-k metal gate stacks. A trapping of the substrate charge into the Si:HfO$_2$ layer and its back tunneling during erase and program pulses, respectively, resulted in a continuous charge transport through the interfacial SiON layer during cycling. This led to its wear-out. Moreover, it could be shown that this charge transport is further enhanced in case of a ferroelectric gate stack. The resulting degradation of the interfacial SiON layer was suggested as the main cause of the endurance deterioration in the Si:HfO$_2$-based MFIS-FET devices. The endurance measurements performed with additional detrapping pulses confirmed that a simple detrapping of electrons trapped during the positive erase pulses cannot improve the endurance properties. Although the endurance characteristics of the studied Si:HfO$_2$-based FeFETs were inferior to those of devices with perovskite ferroelectric materials, which can withstand up to 10^{12} cycles [105], they still were able to meet the modern requirements of the Flash memories (10^4 – 10^5 program/erase cycles [4]).

A detailed study of the charge trapping in the Si:HfO$_2$-based MFIS-FeFET devices and its impact on the memory operation was performed in this work. Besides aggravating the endurance properties, as discussed above, the charge trapping was shown to impair a fast erase of the memory cells. An experimental evidence of a strong electron trapping at typical erase voltages of 3.5 – 5 V was obtained by implementing a single-pulse technique. The estimated characteristic trapping times lay in the range of sub-nanosecond. This makes a complete elimination of trapping during the erase operation impossible for the studied gate stacks. Directly after the erase pulse the ferroelectric memory window was masked by the charge of the trapped electrons. Since the V_{TH} value in the erased state was altered by the trapped electrons, its readout could be performed only after the detrapping processes have set in. At zero volts this process was shown to require several milliseconds. Therefore, the effective erase time is also reduced to milliseconds. A modified approach for the erase operation was proposed in this work in order to mitigate the impact of trapping and increase the effective erase speed. The modified erase pulse consisted of two consecutive pulses – a positive pulse, which induced a positive ferroelectric polarisation, and a negative pulse, which accelerated the detrapping of captured electrons. This modified erase operation enabled to reduce the effective erase time to nanosecond range.

The impact of scaling of Si:HfO$_2$-based MFIS-FETs down to the gate length of 28 nm on their memory performance was investigated. The scaled devices demonstrated characteristics comparable to that of the long channel structures: program and erase times in the range of several nanoseconds (down to 10 ns) with voltages of 4 – 6 V, endurance capability up to 10^4 cycles and a comparable residual *MW* of 0.8 V projected after 10 years at room temperature. The detected differences in the behaviour between the long and short channel devices, such as shift of operation voltages and altered retention behaviour, could be, for the most part, attributed to transistor short channel effects (here V_{TH} roll-off). Therefore, a careful adjustment of the channel implant profiles in scaled cells is expected to provide behaviour similar to the long-channel devices.

In summary, the key memory characteristics of the Si:HfO$_2$-based FeFET devices, studied in this work, are shown in Table 9.1 in comparison to the state-of-the-art ferroelectric transistors with perovskite ferroelectric materials and current NVM technology, the floating-gate. Si:HfO$_2$-based memory cells demonstrated superior properties in respect to operation speed and voltages and comparable retention behaviour. Moreover, they outperform the scaling potential of the devices with SBT ferroelectric films. The capability of the HfO_2-based ferroelectrics to be integrated into the 3D structures, shown in [140], provides the potential to fabricate HfO_2-based FeFETs in non-planar configurations, such as FinFET and 3D array architectures [33], [9], and, thus, to continue the FeFET scaling along with the CMOS technology. The main drawbacks of the current $Si:HfO_2$-based FeFET devices, as identified in this work, are the endurance and charge trapping that is superimposed with the ferroelectric switching. Further studies should be performed in order to minimise the parasitic trapping. This is also expected to improve the endurance characteristics. For this purpose the gate stack should be optimized in a way that enables to moderate the electric field across the interface layer, so that the charge injection probability will decrease. Implementation of high-k interfacial layers or HfO_2 films with reduced P_R values may be a possible solution. An important aspect of scaling, which was not considered in this work, is its impact on the uniformity of the properties of single devices. This is especially in case of the polycrystalline ferroelectric films [263]. As the lateral dimensions of the device decreases to an extent presented here, the active area of the ferroelectric gate is only represented by a few grains/ ferroelectric domains. Therefore, the switching properties of each individual grain start to gain impact on the overall device behaviour. Statistical analyses on the single device structures should be performed in order to address this issue.

Table 9.1 Key memory characteristics of the $Si:HfO_2$-based FeFETs, studied in this work, in comparison to the state-of-the-art ferroelectric transistors with SBT ferroelectric films and floating-gate memory cell.

Properties	Floating-gate cells [4]	SBT-based FeFET [112]	Si:HfO$_2$-based FeFET (this work)
P_R ($\mu C/cm^2$)	–	$10 - 24$ [49]	$18 - 24$
E_C (kV/cm)	–	$40 - 60$ [49]	$\sim 800 - 900$
ε	–	300 [49]	$20 - 25$
Operation voltage	$15 - 17$ V	$4 - 7$ V	$4 - 6$ V
Write times	$\mu s - ms$	$\mu s - ms$	$30 - 100$ ns
Endurance (cycles)	$10^4 - 10^5$	10^9	$10^4 - 10^5$
Retention	10 years	10 years	10 years
Current scaling node	16 nm [3]	260 nm	28 nm
Compatibility with CMOS process	+	–	+

Bibliography

[1]   J. E. Brewer and M. Gill, *Nonvolatile Memory Technologies with Emphasis on Flash A comprehensive guide to understanding and using NVM devices*. Hoboken, New Jersey: John Wiley & Sons, 2008.

[2]   G. E. Moore, "Cramming more components onto integrated circuits," *Proceedings of the IEEE*, vol. 86, no. 1, p. 82, 1998.

[3]   The Micron Technology, Inc. Press Release, "Micron Unveils 16-Nanometer Flash Memory Technology," *(http://www.micron.com/about/news-and-events/media-kits/16nm-nandm)*, 16-Jul-2013. .

[4]   *International Technology Roadmap for Semiconductors*. Semiconductor Industry Association (available online at www.itrs.net/reports.htlm), 2012.

[5]   K. Prall, "Scaling non-volatile memory below 30nm," in *Non-Volatile Semiconductor Memory Workshop, 2007 22nd IEEE*, 2007, pp. 5–10.

[6]   S. Lee, "Scaling Challenges in NAND Flash Device toward 10nm Technology," in *Memory Workshop (IMW), 2012 4th IEEE International*, 2012, pp. 1–4.

[7]   S. Hong, "Memory technology trend and future challenges," in *Electron Devices Meeting (IEDM), 2010 IEEE International*, 2010, pp. 12–4.

[8]   E. Maayan, R. Dvir, J. Shor, Y. Polansky, Y. Sofer, I. Bloom, D. Avni, B. Eitan, Z. Cohen, and M. Meyassed, "A 512 Mb NROM flash data storage memory with 8 MB/s data rate," in *Solid-State Circuits Conference, 2002. Digest of Technical Papers. ISSCC. 2002 IEEE International*, 2002, vol. 1, pp. 100–101.

[9]   J. Jang, H.-S. Kim, W. Cho, H. Cho, J. Kim, S. I. Shim, Y. Jang, J.-H. Jeong, B.-K. Son, and D. W. Kim, "Vertical cell array using TCAT (Terabit Cell Array Transistor) technology for ultra high density NAND flash memory," in *VLSI Technology, 2009 Symposium on*, 2009, pp. 192–193.

[10]   T. Kawahara, K. Ito, R. Takemura, and H. Ohno, "Spin-transfer torque RAM technology: Review and prospect," *Microelectronics Reliability*, vol. 52, no. 4, pp. 613–627, Apr. 2012.

[11]   H.-S. P. Wong, S. Raoux, S. Kim, J. Liang, J. P. Reifenberg, B. Rajendran, M. Asheghi, and K. E. Goodson, "Phase Change Memory," *Proceedings of the IEEE*, vol. 98, no. 12, pp. 2201–2227, Dec. 2010.

[12]   D. S. Jeong, R. Thomas, R. S. Katiyar, J. F. Scott, H. Kohlstedt, A. Petraru, and C. S. Hwang, "Emerging memories: resistive switching mechanisms and current status," *Reports on Progress in Physics*, vol. 75, no. 7, p. 076502, Jul. 2012.

[13] J. Hutchby and M. Garner, "Assessment of the Potential and Maturity of Selected Emerging Research Memory Technologies," in *Workshop and ERD/ERM Working Group Meeting, www.itrs.net*, 2010, pp. 1–55.

[14] S. Sakai, M. Takahashia, K. Takeuchib, Q.-H. Lia, T. Horiuchia, S. Wanga, K.-Y. Yuna, M. Takamiyac, and T. Sakuraic, "Highly scalable Fe (Ferroelectric)-NAND cell with MFIS (Metal-Ferroelectric-Insulator-Semiconductor) structure for sub-10nm tera-bit capacity NAND flash memories," in *Non-Volatile Semiconductor Memory Workshop/ International Conference on Memory Technology and Design. NVSMW/ICMTD 2008. Joint*, 2008, pp. 103–105.

[15] I. M. Ross, "Semiconductor Translating Device," US Patent No 2,791,760, 1957.

[16] N. Setter, D. Damjanovic, L. Eng, G. Fox, S. Gevorgian, S. Hong, A. Kingon, H. Kohlstedt, N. Y. Park, G. B. Stephenson, I. Stolitchnov, A. K. Tagantsev, D. V. Taylor, T. Yamada, and S. Streiffer, "Ferroelectric thin films: Review of materials, properties, and applications," *Journal of Applied Physics*, vol. 100, no. 5, p. 051606, 2006.

[17] R. Waser, *Nanoelectronics and Information Technology: Advanced Electronic Materials and Novel Devices*. Weinheim: WILEY-VCH Verlag GmbH and Co. KGaA, 2005.

[18] T. S. Böscke, J. Müller, D. Bräuhaus, U. Schröder, and U. Böttger, "Ferroelectricity in hafnium oxide thin films," *Applied Physics Letters*, vol. 99, no. 10, pp. 102903–102903, 2011.

[19] J. Müller, T. S. Böscke, D. Bräuhaus, U. Schröder, U. Böttger, J. Sundqvist, P. Kücher, T. Mikolajick, and L. Frey, "Ferroelectric $Zr_{0.5}Hf_{0.5}O_2$ thin films for nonvolatile memory applications," *Applied Physics Letters*, vol. 99, no. 11, pp. 112901–112901, 2011.

[20] J. Müller, U. Schröder, T. S. Böscke, I. Müller, U. Böttger, L. Wilde, J. Sundqvist, M. Lemberger, P. Kücher, T. Mikolajick, and L. Frey, "Ferroelectricity in yttrium-doped hafnium oxide," *Journal of Applied Physics*, vol. 110, no. 11, p. 114113, 2011.

[21] *International Technology Roadmap for Semiconductors*. Semiconductor Industry Association (available online at www.itrs.net/reports.htlm), 2013.

[22] T. S. Böscke, "Crystalline Hafnia and Zirconia based Dielectrics for Memory Applications," Doctoral dissertation, Technische Universität Hamburg-Harburg, 2010.

[23] E. Yurchuk, J. Müller, S. Knebel, J. Sundqvist, A. P. Graham, T. Melde, U. Schröder, and T. Mikolajick, "Impact of layer thickness on the ferroelectric behaviour of silicon doped hafnium oxide thin films," *Thin Solid Films*, vol. 533, pp. 88–92, Apr. 2013.

[24] J. Müller, E. Yurchuk, T. Schlösser, J. Paul, R. Hoffmann, S. Müller, D. Martin, S. Slesazeck, P. Polakowski, and J. Sundqvist, "Ferroelectricity in HfO_2 enables

nonvolatile data storage in 28 nm HKMG," in *VLSI Technology (VLSIT), Symposium on*, 2012, pp. 25–26.

[25] S. M. Sze and K. N. Kwok, *Physics of Semiconductor Devices*, 3rd Edition. Hoboken, New Jersey: John Wiley & Sons, 2007.

[26] D. K. Schroder, *Semiconductor Materials and Device Characterisation*, 3rd Edition. Hoboken, New Jersey: John Wiley & Sons, 2006.

[27] P. Cappelleti, C. Golla, P. Olivo, and E. Zanoni, *Flash Memories*. Boston, Dordrecht, Londen: Kluwer academic publishers, 2000.

[28] R. Sbiaa, H. Meng, and S. N. Piramanayagam, "Materials with perpendicular magnetic anisotropy for magnetic random access memory," *physica status solidi (RRL) - Rapid Research Letters*, vol. 5, no. 12, pp. 413–419, Dec. 2011.

[29] A. Sawa, "Resistive switching in transition metal oxides," *Materials today*, vol. 11, no. 6, pp. 28–36, 2008.

[30] H.-T. Lue, S.-Y. Wang, E.-K. Lai, Y.-H. Shih, S.-C. Lai, L.-W. Yang, K.-C. Chen, J. Ku, K.-Y. Hsieh, and R. Liu, "BE-SONOS: A bandgap engineered SONOS with excellent performance and reliability," in *Electron Devices Meeting, 2005. IEDM Technical Digest. IEEE International*, 2005, pp. 547–550.

[31] Y. Shin, J. Choi, C. Kang, C. Lee, K.-T. Park, J.-S. Lee, J. Sel, V. Kim, B. Choi, and J. Sim, "A novel NAND-type MONOS memory using 63nm process technology for multi-gigabit flash EEPROMs," in *Electron Devices Meeting, 2005. IEDM Technical Digest. IEEE International*, 2005, pp. 327–330.

[32] Y. Polansky, A. Lavan, R. Sahar, O. Dadashev, Y. Betser, G. Cohen, E. Maayan, B. Eitan, F.-L. Ni, and Y.-H. J. Ku, "A 4b/cell NROM 1Gb Data-Storage Memory," in *Solid-State Circuits Conference, 2006. ISSCC 2006. Digest of Technical Papers. IEEE International*, 2006, pp. 448–458.

[33] R. Katsumata, M. Kito, Y. Fukuzumi, M. Kido, H. Tanaka, Y. Komori, M. Ishiduki, J. Matsunami, T. Fujiwara, and Y. Nagata, "Pipe-shaped BiCS flash memory with 16 stacked layers and multi-level-cell operation for ultra high density storage devices," in *VLSI Technology, 2009 Symposium on*, 2009, pp. 136–137.

[34] T. Miyazaki and N. Tezuka, "Giant magnetic tunneling effect in $Fe/Al_2O_3/Fe$ junction," *Journal of Magnetism and Magnetic Materials*, vol. 139, no. 3, pp. L231–L234, 1995.

[35] Jian-Gang Zhu, "Magnetoresistive Random Access Memory: The Path to Competitiveness and Scalability," *Proceedings of the IEEE*, vol. 96, no. 11, pp. 1786–1798, Nov. 2008.

[36] B. C. Lee, P. Zhou, J. Yang, Y. Zhang, B. Zhao, E. Ipek, O. Mutlu, and D. Burger, "Phase-change technology and the future of main memory," *Micro, IEEE*, vol. 30, no. 1, pp. 143–143, 2010.

[37]   R. Waser, R. Dittmann, G. Staikov, and K. Szot, "Redox-Based Resistive Switching Memories - Nanoionic Mechanisms, Prospects, and Challenges," *Advanced Materials*, vol. 21, no. 25–26, pp. 2632–2663, Jul. 2009.

[38]   V. V. Zhirnov, R. Meade, R. K. Cavin, and G. Sandhu, "Scaling limits of resistive memories," *Nanotechnology*, vol. 22, no. 25, p. 254027, Jun. 2011.

[39]   J. J. Yang, M.-X. Zhang, J. P. Strachan, F. Miao, M. D. Pickett, R. D. Kelley, G. Medeiros-Ribeiro, and R. S. Williams, "High switching endurance in TaO_x memristive devices," *Applied Physics Letters*, vol. 97, no. 23, p. 232102, 2010.

[40]   M. N. Kozicki, M. Park, and M. Mitkova, "Nanoscale Memory Elements Based on Solid-State Electrolytes," *IEEE Transactions On Nanotechnology*, vol. 4, no. 3, pp. 331–338, May 2005.

[41]   G. Panomsuwan, O. Takai, and N. Saito, "Enhanced memory window of $Au/BaTiO_3/SrTiO_3/Si(001)$ MFIS structure with high c-axis orientation for non-volatile memory applications," *Applied Physics A*, vol. 108, no. 2, pp. 337–342, Aug. 2012.

[42]   S. R. Summerfelt, T. S. Moise, K. R. Udayakumar, K. Boku, K. Remack, J. Rodriguez, J. Gertas, H. McAdams, S. Madan, and J. Eliason, "High-Density 8Mb 1T-1C Ferroelectric Random Access Memory Embedded Within a Low-Power 130nm Logic Process," in *Applications of Ferroelectrics, 2007. ISAF 2007. Sixteenth IEEE International Symposium on*, 2007, pp. 9–10.

[43]   K. Fujimoto, T. Kawano, A. Onoe, M. Tamura, M. Umeda, and M. Toda, "1 Tbit/inch^2 very high-density recording in polycrystalline $PZT/SRO/SiO_2/Si$ thin film," in *Applications of Ferroelectrics, 2008. ISAF 2008. 17th IEEE International Symposium on the*, 2008, vol. 2, pp. 1–2.

[44]   J. T. Evans and J. R. I. Suizu, "Static FRAM: An Emerging Nonvolatile Memory Technology," in *Nonvolatile Memory Technology Conference, 7th Biennial IEEE*, Albuquerque, NM, 1998, p. 26.

[45]   L. Goux, G. Russo, N. Menou, J. G. Lisoni, M. Schwitters, V. Paraschiv, D. Maes, C. Artoni, G. Corallo, L. Haspeslagh, D. J. Wouters, R. Zambrano, and C. Muller, "A Highly Reliable 3-D Integrated SBT Ferroelectric Capacitor Enabling FeRAM Scaling," *IEEE Transactions on Electron Devices*, vol. 52, no. 4, pp. 447–453, Apr. 2005.

[46]   Y. J. Park, I. Bae, S. J. Kang, J. Chang, and C. Park, "Control of thin ferroelectric polymer films for non-volatile memory applications," *Dielectrics and Electrical Insulation, IEEE Transactions on*, vol. 17, no. 4, pp. 1135–1163, 2010.

[47]   R. C. G. Naber, K. Asadi, P. W. M. Blom, D. M. de Leeuw, and B. de Boer, "Organic Nonvolatile Memory Devices Based on Ferroelectricity," *Advanced Materials*, vol. 22, no. 9, pp. 933–945, Mar. 2010.

[48]   F. Jona and G. Shirane, *Ferroelectric Crystals*, vol. 1. Pergamon press, 1962.

[49]   H. Ishiwara and M. Okuyama, *Ferroelectric Random Access Memories Fundamentals and Applications*, vol. 9. Springer Verlag, 2007.

[50]   O. Auciello, "A critical comparative review of PZT and SBT-based science and technology for non-volatile ferroelectric memories," *Integrated Ferroelectrics*, vol. 15, no. 1–4, pp. 211–220, Feb. 1997.

[51]   Q.-D. Ling, D.-J. Liaw, C. Zhu, D. S.-H. Chan, E.-T. Kang, and K.-G. Neoh, "Polymer electronic memories: Materials, devices and mechanisms," *Progress in Polymer Science*, vol. 33, no. 10, pp. 917–978, Oct. 2008.

[52]   B. H. Park, B. S. Kang, S. D. Bu, T. W. Noh, J. Lee, and W. Jo, "Lanthanum-substituted bismuth titanate for use in non-volatile memories," *Nature*, vol. 401, pp. 682–684, 1999.

[53]   J. F. Scott, *Ferroelectric Memories*. Springer Verlag, 2000.

[54]   M. Foeth, A. Sfera, P. Stadelmann, and P.-A. Buffat, "A comparison of HREM and weak beam transmission electron microscopy for the quantitative measurement of the thickness of ferroelectric domain walls," *Journal of Electron Microscopy*, vol. 48, no. 6, pp. 717–723, 1999.

[55]   S. Stemmer, S. K. Streiffer, F. Ernst, and M. Rüuhle, "Atomistic structure of 90° domain walls in ferroelectric PbTiO$_3$ thin films," *Philosophical Magazine A*, vol. 71, no. 3, pp. 713–724, Mar. 1995.

[56]   D. Damjanovic, "Hysteresis in piezoelectric and ferroelectric materials," *The science of hysteresis*, vol. 3, pp. 337–465, 2005.

[57]   H. M. Duiker, P. D. Beale, J. F. Scott, C. A. Paz de Araujo, B. M. Melnick, J. D. Cuchiaro, and L. D. McMillan, "Fatigue and switching in ferroelectric memories: Theory and experiment," *Journal of Applied Physics*, vol. 68, no. 11, p. 5783, 1990.

[58]   I. Stolichnov, A. Tagantsev, N. Setter, J. S. Cross, and M. Tsukada, "Crossover between nucleation-controlled kinetics and domain wall motion kinetics of polarization reversal in ferroelectric films," *Applied Physics Letters*, vol. 83, no. 16, p. 3362, 2003.

[59]   Y. Ishibashi and Y. Tagaki, "Note on ferroelectric domain switching," *Journal of the Physical Society of Japan*, vol. 31, no. 2, p. 506, 1971.

[60]   J. F. Scott, L. Kammerdiner, M. Parris, S. Traynor, V. Ottenbacher, A. Shawabkeh, and W. F. Oliver, "Switching kinetics of lead zirconate titanate submicron thin-film memories," *Journal of Applied Physics*, vol. 64, no. 2, p. 787, 1988.

[61]   A. Tagantsev, I. Stolichnov, N. Setter, J. Cross, and M. Tsukada, "Non-Kolmogorov-Avrami switching kinetics in ferroelectric thin films," *Physical Review B*, vol. 66, no. 21, Dec. 2002.

[62]   O. Lohse, M. Grossmann, U. Boettger, D. Bolten, and R. Waser, "Relaxation mechanism of ferroelectric switching in Pb(Zr,Ti)O$_3$ thin films," *Journal of Applied Physics*, vol. 89, no. 4, p. 2332, 2001.

[63]   Y. W. So, D. J. Kim, T. W. Noh, J.-G. Yoon, and T. K. Song, "Polarization-switching mechanisms for epitaxial ferroelectric Pb(Zr, Ti)O$_3$ films," *Journal of the Korean Physical Society*, vol. 46, no. 1, pp. 40–43, 2005.

[64]   I. Stolichnov, L. Malin, E. Colla, A. K. Tagantsev, and N. Setter, "Microscopic aspects of the region-by-region polarization reversal kinetics of polycrystalline ferroelectric Pb(Zr,Ti)O$_3$ films," *Applied Physics Letters*, vol. 86, no. 1, p. 012902, 2005.

[65]   D. J. Wouters, R. Nouwen, G. J. Norga, A. Bartic, L. Van Poucke, and H. E. Maes, "Switching quality of thin-film PZT ferroelectric capacitors," *Le Journal de Physique IV*, vol. 08, no. PR9, pp. Pr9–205–Pr9–208, Dec. 1998.

[66]   P. K. Larsen, G. L. M. Kampschöer, M. J. E. Ulenaers, G. A. C. M. Spierings, and R. Cuppens, "Nanosecond switching of thin ferroelectric films," *Applied Physics Letters*, vol. 59, no. 5, p. 611, 1991.

[67]   P. C. Joshi and S. B. Krupanidhi, "Switching, fatigue, and retention in ferroelectric Bi$_4$Ti$_3$O$_{12}$ thin films," *Applied Physics Letters*, vol. 62, no. 16, p. 1928, 1993.

[68]   J. F. Scott, "Ferroelectrics go bananas," *Journal of Physics Condensed Matter*, vol. 20, no. 2, p. 21001, 2008.

[69]   B. Jaffe, W. R. Cook, and H. Jaffe, *Piezoelectric Ceramics*. London: Academic press, 1971.

[70]   L. Pintilie and M. Alexe, "Ferroelectric-like hysteresis loop in nonferroelectric systems," *Applied Physics Letters*, vol. 87, no. 11, pp. 112903–112903, 2005.

[71]   M. E. Lines and A. M. Grass, *Principles and Applications of Ferroelectrics and Related Materials*. Oxford: Clarendon Press, 1977.

[72]   D. Bolten, O. Lohse, M. Grossmann, and R. Waser, "Reversible and irreversible domain wall contributions to the polarization in ferroelectric thin films," *Ferroelectrics*, vol. 221, no. 1, pp. 251–257, Jan. 1999.

[73]   C. J. Brennan, "Characterization and modelling of thin-film ferroelectric capacitors using C-V analysis," *Integrated Ferroelectrics*, vol. 2, no. 1–4, pp. 73–82, Nov. 1992.

[74]   G. Shirane, E. Sawaguchi, and Y. Takagi, "Dielectric properties of lead zirconate," *Physical Review*, vol. 84, no. 3, p. 476, 1951.

[75]   S.-E. Park, M.-J. Pan, K. Markowski, S. Yoshikawa, and L. E. Cross, "Electric field induced phase transition of antiferroelectric lead lanthanum zirconate titanate stannate ceramics," *Journal of applied physics*, vol. 82, no. 4, pp. 1798–1803, 1997.

[76]   A. Moriellis, "Piezoresponse force microscopy of ferroelectric thin films," Doctoral dissertation, University of Groningen, 2009.

[77]   E. Fatuzzo and W. J. Merz, *Ferroelectricity*, vol. 7, 7 vols. Amsterdam: Noth-Holland Publishing company, 1967.

[78]   A. K. Tagantsev, I. Stolichnov, E. L. Colla, and N. Setter, "Polarization fatigue in ferroelectric films: Basic experimental findings, phenomenological scenarios, and microscopic features," *Journal of Applied Physics*, vol. 90, no. 3, p. 1387, 2001.

[79]   M. Dawber and J. F. Scott, "A model for fatigue in ferroelectric perovskite thin films," *Applied Physics Letters*, vol. 76, no. 8, p. 1060, 2000.

[80]   E. L. Colla, D. V. Taylor, A. K. Tagantsev, and N. Setter, "Discrimination between bulk and interface scenarios for the suppression of the switchable polarization (fatigue) in $Pb(Zr,Ti)O_3$ thin films capacitors with Pt electrodes," *Applied Physics Letters*, vol. 72, no. 19, p. 2478, 1998.

[81]   D. Dimos, W. L. Warren, and H. N. Al-Shareef, "Degradation mechanisms and reliability issues for ferroelectric thin films," in *Thin Film Ferroelectric Materials and Devices/ edited by R. Ramesh*, Boston, Dordrecht, Londen: Kluwer academic publishers, 1997, pp. 199–219.

[82]   D. Dimos, W. L. Warren, M. B. Sinclair, B. A. Tuttle, and R. W. Schwartz, "Photoinduced hysteresis changes and optical storage in $(Pb,La)(Zr,Ti)O_3$ thin films and ceramics," *Journal of Applied Physics*, vol. 76, no. 7, p. 4305, 1994.

[83]   W. L. Warren, D. Dimos, B. A. Tuttle, G. E. Pike, R. W. Schwartz, P. J. Clews, and D. C. McIntyre, "Polarization suppression in $Pb(Zr,Ti)O_3$ thin films," *Journal of Applied Physics*, vol. 77, no. 12, p. 6695, 1995.

[84]   S. M. Yang, T. H. Kim, J.-G. Yoon, and T. W. Noh, "Nanoscale Observation of Time-Dependent Domain Wall Pinning as the Origin of Polarization Fatigue," *Advanced Functional Materials*, vol. 22, no. 11, pp. 2310–2317, Jun. 2012.

[85]   J. F. Scott and M. Dawber, "Oxygen-vacancy ordering as a fatigue mechanism in perovskite ferroelectrics," *Applied Physics Letters*, vol. 76, no. 25, p. 3801, 2000.

[86]   S. B. Desu and I. K. Yoo, "Electrochemical models of failure in oxide perovskites," *Integrated Ferroelectrics*, vol. 3, no. 4, pp. 365–376, Dec. 1993.

[87]   I. Stolichnov, A. K. Tagantsev, E. L. Colla, and N. Setter, "Cold-field-emission test of the fatigued state of $Pb(Zr_xTi_{1-x})O_3$ films," *Applied Physics Letters*, vol. 73, no. 10, p. 1361, 1998.

[88]   X. Du and I.-W. Chen, "Fatigue of $Pb(Zr_{0.53}Ti_{0.47})O_3$ ferroelectric thin films," *Journal of Applied Physics*, vol. 83, no. 12, p. 7789, 1998.

[89]   T. Mihara, H. Watanabe, and C. A. Paz De Araujo, "Polarisation fatigue characteristics of sol-gel ferroelectric $Pb(Zr_{0.4}Ti_{0.6})O_3$ thin-film capacitors," *Japanese Journal of Applied Physics*, vol. 33, no. 7A, pp. 3996–4002, 1994.

[90]   S. Sun and P. A. Fuierer, "Modeling of depolarization in ferroelectric thin films," *Integrated Ferroelectrics*, vol. 23, no. 1–4, pp. 45–64, Jul. 1999.

[91] R. R. Mehta, "Depolarization fields in thin ferroelectric films," *Journal of Applied Physics*, vol. 44, no. 8, p. 3379, 1973.

[92] M. Takahashi, H. Sugiyama, T. Nakaiso, K. Kodama, M. Noda, and M. Okuyama, "Analyses and improvement of retention time of memorized state of Metal-Ferroelectric-Insulator-Semiconductor structure for Ferroelectric Gate FET memory," *Japanese Journal of Applied Physics*, vol. 40, no. 4B, pp. 2923–2927, 2001.

[93] A. K. Tagantsev, M. Landivar, E. Colla, and N. Setter, "Identification of passive layer in ferroelectric thin films from their switching parameters," *Journal of Applied Physics*, vol. 78, no. 4, p. 2623, 1995.

[94] T. Hatanaka, R. Yajima, T. Horiuchi, S. Wang, X. Zhang, M. Takahashi, S. Sakai, and K. Takeuchi, "Ferroelectric (Fe)-NAND flash memory with non-volatile page buffer for data center application enterprise solid-state drives (SSD)," in *VLSI Circuits, 2009 Symposium on*, 2009, pp. 78–79.

[95] Y. K. Hong, D. J. Jung, S. K. Kang, H. S. Kim, J. Y. Jung, H. K. Koh, J. H. Park, D. Y. Choi, S. E. Kim, and W. S. Ann, "130 nm-technology, 0.25 μm^2, 1T1C FRAM Cell for SoC (System-on-a-Chip)-friendly Applications," in *VLSI Technology, 2007 IEEE Symposium on*, 2007, pp. 230–231.

[96] K. Yamaoka, S. Iwanari, Y. Marakuki, H. Hirano, M. Sakagami, T. Nakakuma, T. Miki, and Y. Gohou, "A 0.9V 1T1C SBT-Based Embedded Non-Volatile FeRAM with a Reference Voltage Scheme and Multi-Layer Shielded Bit-Line Structure," presented at the IEEE International Solid-State Circuits Conference, 2004.

[97] H. Kohlstedt, Y. Mustafa, A. Gerber, A. Petraru, M. Fitsilis, R. Meyer, U. Böttger, and R. Waser, "Current status and challenges of ferroelectric memory devices," *Microelectronic Engineering*, vol. 80, pp. 296–304, Jun. 2005.

[98] T. Mikolajick, C. Dehm, W. Hartner, I. Kasko, M. J. Kastner, N. Nagel, M. Moert, and C. Mazure, "FeRAM technology for high density applications," *Microelectronics Reliability*, vol. 41, no. 7, pp. 947–950, 2001.

[99] M. Qazi, M. Clinton, S. Bartling, and A. P. Chandrakasan, "A Low-Voltage 1 Mb FRAM in 0.13 μm CMOS Featuring Time-to-Digital Sensing for Expanded Operating Margin," *IEEE Journal of Solid-State Circuits*, vol. 47, no. 1, pp. 141–150, Jan. 2012.

[100] R. Zambrano, "Challenges for Integration of Embedded FeRAMs in the sub-180 nm Regime," *Integrated ferroelectrics*, vol. 53, no. 1, pp. 247–255, 2003.

[101] C. Y. Wu, "A new Ferroelectric Memory Device, Metal-Ferroelectric-Semiconductor Transistor," *IEEE Transaction on Electron Devices*, vol. ED-21, p. 490, 1974.

[102] S. Y. Wu, "Memory retention and switching behavior of metal-ferroelectric-semiconductor transistors," *Ferroelectrics*, vol. 11, no. 1, pp. 379–383, Jan. 1976.

[103] T. Hirai, K. Teramoto, K. Nagashima, H. Koike, and Y. Tarui, "Characterization of Metal/Ferroelectric/Insulator/Semiconductor Structure with CeO_2 Buffer layer," *Japanese Journal of Applied Physics*, vol. 34, pp. 4163–4166, 1995.

[104] B.-E. Park, S. Shouriki, E. Tokumitsu, and H. Ishiwara, "Fabrication of $PbZr_xTi_{1-x}O_3$ Films on Si structures using Y_2O_3 buffer layers," *Japanese Journal of Applied Physics*, vol. 37, pp. 5145–5149, 1998.

[105] S. Sakai and R. Ilangovan, "Metal-Ferroelectric-Insulator-Semiconductor Memory FET With Long Retention and High Endurance," *IEEE Electron Device Letters*, vol. 25, no. 6, pp. 369–371, Jun. 2004.

[106] M. Takahashi, K. Aizawa, B.-E. Park, and H. Ishiwara, "Thirty-day-long data retention in Ferroelectric -Gate Field Effect Transistor with HfO_2 buffer layer," *Japanese Journal of Applied Physics*, vol. 44, no. 8, pp. 6218–6220, 2005.

[107] M. Ullmann, *Ferroelektrische Feldeffekttransistoren: Modellierung und Anwendung*, vol. 354. Düsseldorf: VDI Verlag, 2002.

[108] Hang-Ting Lue, Chien-Jang Wu, and Tseung-Yuen Tseng, "Device modeling of ferroelectric memory field-effect transistor (FeMFET)," *IEEE Transactions on Electron Devices*, vol. 49, no. 10, pp. 1790–1798, 2002.

[109] Y. Watanabe, "Theoretical stability of the polarization in insulating ferroelectric/semiconductor structures," *Journal of Applied Physics*, vol. 83, no. 4, p. 2179, 1998.

[110] T. P. Ma and J.-P. Han, "Why is nonvolatile ferroelectric memory field-effect transistor still elusive?," *Electron Device Letters, IEEE*, vol. 23, no. 7, pp. 386–388, 2002.

[111] S. Sakai and M. Takahashi, "Recent Progress of Ferroelectric-Gate Field-Effect Transistors and Applications to Nonvolatile Logic and FeNAND Flash Memory," *Materials*, vol. 3, no. 11, pp. 4950–4964, 2010.

[112] L. V. Hai, M. Takahashi, and S. Sakai, "Downsizing of Ferroelectric-Gate Field-Effect-Transistors for Ferroelectric-NAND Flash Memory Cells," in *Memory Workshop (IMW), 3rd IEEE International*, 2011, pp. 1–4.

[113] Y. Nakao, T. Nakamura, A. Kamisawa, and H. Takasu, "Study on ferroelectric thin films for application to NDRO non-volatile memories," *Integrated Ferroelectrics*, vol. 6, no. 1–4, pp. 23–34, Jan. 1995.

[114] E. Tokumitsu, G. Fujii, and H. Ishiwara, "Electrical properties of Metal-Ferroelectric-Insulator-Semiconductor (MFIS) and Metal-Ferroelectric-Metal-Insulator-Semiconductor (MFMIS)-FETs Using Ferroelectric $SrBi_2Ta_2O_9$ Film and $SrTa_2O_6$/SiON Buffer Layer," *Japanese Journal of Applied Physics*, vol. 39, pp. 2125–2130, 2000.

[115] H. Ishiwara, "Current Status and Prospects of FET-type Ferroelectric Memories," *JOURNAL OF SEMICONDUCTOR TECHNOLOGY AND SCIENCE*, vol. 1, pp. 1–14, 2001.

[116] J. Hoffman, X. Pan, J. W. Reiner, F. J. Walker, J. P. Han, C. H. Ahn, and T. P. Ma, "Ferroelectric Field Effect Transistors for Memory Applications," *Advanced Materials*, vol. 22, no. 26–27, pp. 2957–2961, Apr. 2010.

[117] J. Wang, H. P. Li, and R. Stevens, "Hafnia and hafnia-toughened ceramics," *Journal of materials science*, vol. 27, no. 20, pp. 5397–5430, 1992.

[118] O. Ohtaka, H. Fukui, T. Kunisada, T. Fujisawa, K. Funakoshi, W. Utsumi, T. Irifune, K. Kuroda, and T. Kikegawa, "Phase relations and volume changes of hafnia under high pressure and high temperature," *Journal of the American Ceramic Society*, vol. 84, no. 6, pp. 1369–1373, 2001.

[119] T. S. Böscke, S. Teichert, D. Bräuhaus, J. Müller, U. Schröder, U. Böttger, and T. Mikolajick, "Phase transitions in ferroelectric silicon doped hafnium oxide," *Applied Physics Letters*, vol. 99, no. 11, p. 112904, 2011.

[120] S. V. Ushakov, A. Navrotsky, Y. Yang, S. Stemmer, K. Kukli, M. Ritala, M. A. Leskelä, P. Fejes, A. Demkov, C. Wang, B.-Y. Nguyen, D. Triyoso, and P. Tobin, "Crystallization in hafnia- and zirconia-based systems," *physica status solidi (b)*, vol. 241, no. 10, pp. 2268–2278, Aug. 2004.

[121] A. Navrotsky, "Thermochemical insights into refractory ceramic materials based on oxides with large tetravalent cations," *Journal of Materials Chemistry*, vol. 15, no. 19, p. 1883, 2005.

[122] C.-K. Lee, E. Cho, H.-S. Lee, C. Hwang, and S. Han, "First-principles study on doping and phase stability of HfO_2," *Physical Review B*, vol. 78, no. 1, Jul. 2008.

[123] K. Tomida, K. Kita, and A. Toriumi, "Dielectric constant enhancement due to Si incorporation into HfO_2," *Applied Physics Letters*, vol. 89, no. 14, p. 142902, 2006.

[124] K. Kita, K. Kyuno, and A. Toriumi, "Permittivity increase of yttrium-doped HfO_2 through structural phase transformation," *Applied Physics Letters*, vol. 86, no. 10, p. 102906, 2005.

[125] T. S. Böscke, P. Y. Hung, P. D. Kirsch, M. A. Quevedo-Lopez, and R. Ramírez-Bon, "Increasing permittivity in HfZrO thin films by surface manipulation," *Applied Physics Letters*, vol. 95, no. 5, p. 052904, 2009.

[126] S. Govindarajan, T. S. Böscke, P. Sivasubramani, P. D. Kirsch, B. H. Lee, H.-H. Tseng, R. Jammy, U. Schröder, S. Ramanathan, and B. E. Gnade, "Higher permittivity rare earth doped HfO_2 for sub-45-nm metal-insulator-semiconductor devices," *Applied Physics Letters*, vol. 91, no. 6, p. 062906, 2007.

[127] G. Pant, A. Gnade, M. J. Kim, R. M. Wallace, B. E. Gnade, M. A. Quevedo-Lopez, and P. D. Kirsch, "Effect of thickness on the crystallization of ultrathin HfSiON gate dielectrics," *Applied Physics Letters*, vol. 88, no. 3, p. 032901, 2006.

[128] D. H. Triyoso, P. J. Tobin, B. E. White, R. Gregory, and X. D. Wang, "Impact of film properties of atomic layer deposited HfO$_2$ resulting from annealing with a TiN capping layer," *Applied Physics Letters*, vol. 89, no. 13, p. 132903, 2006.

[129] R. B. van Dover, M. L. Green, L. Manchanda, L. F. Schneemeyer, and T. Siegrist, "Composition-dependent crystallization of alternative gate dielectrics," *Applied Physics Letters*, vol. 83, no. 7, p. 1459, 2003.

[130] S. Mueller, J. Mueller, A. Singh, S. Riedel, J. Sundqvist, U. Schroeder, and T. Mikolajick, "Incipient Ferroelectricity in Al-Doped HfO$_2$ Thin Films," *Advanced Functional Materials*, vol. 22, no. 11, pp. 2412–2417, Jun. 2012.

[131] U. Schroeder, S. Mueller, J. Mueller, E. Yurchuk, D. Martin, C. Adelmann, T. Schloesser, R. van Bentum, and T. Mikolajick, "Hafnium Oxide Based CMOS Compatible Ferroelectric Materials," *ECS Journal of Solid State Science and Technology*, vol. 2, no. 4, pp. N69–N72, 2013.

[132] J. Müller, T. S. Böscke, U. Schröder, S. Mueller, D. Bräuhaus, U. Böttger, L. Frey, and T. Mikolajick, "Ferroelectricity in Simple Binary ZrO$_2$ and HfO$_2$," *Nano Letters*, vol. 12, no. 8, pp. 4318–4323, Aug. 2012.

[133] T. Olsen, U. Schröder, S. Müller, A. Krause, D. Martin, A. Singh, J. Müller, M. Geidel, and T. Mikolajick, "Co-sputtering yttrium into hafnium oxide thin films to produce ferroelectric properties," *Applied Physics Letters*, vol. 101, no. 8, p. 082905, 2012.

[134] D. B. Marshall, M. R. Jarnes, and J. R. Porter, "Structural and Mechanical Property Changes in Toughened Magnesia-Partially-Stabilized Zirconia at Low Temperatures," *Journal of the American Ceramic Society*, vol. 72, no. 2, pp. 218–227, 1989.

[135] E. H. Kisi, C. J. Howard, and R. J. Hill, "Crystal structure of orthorhombic zirconia in partially stabilized zirconia," *Journal of the American Ceramic Society*, vol. 72, no. 9, pp. 1757–1760, 1989.

[136] E. H. Kisi, "Influence of Hydrostatic Pressure on the t→ o Transformation in Mg-PSZ Studied by In Situ Neutron Diffraction," *Journal of the American Ceramic Society*, vol. 81, no. 3, pp. 741–745, 1998.

[137] K. Mistry, C. Allen, C. Auth, B. Beattie, D. Bergstrom, M. Bost, M. Brazier, M. Buehler, A. Cappellani, and R. Chau, "A 45nm logic technology with high-k+metal gate transistors, strained silicon, 9 Cu interconnect layers, 193nm dry patterning, and 100% Pb-free packaging," in *Electron Devices Meeting, 2007. IEDM 2007. IEEE International*, 2007, pp. 247–250.

[138] K. Choi, T. Ando, E. A. Cartier, A. Kerber, V. Paruchuri, J. Iacoponi, and V. Narayanan, "(Invited) The Past, Present and Future of High-*k*/Metal Gates," *ECS Transactions*, vol. 53, no. 3, pp. 17–26, May 2013.

[139] J. Yuan, C. Gruensfelder, K. Y. Lim, T. Wallner, M. K. Jung, M. J. Sherony, Y. M. Lee, J. Chen, C. W. Lai, and Y. T. Chow, "Performance elements for 28nm gate length bulk devices with gate first high-*k* metal gate," in *Solid-State and Integrated Circuit Technology (ICSICT), 2010 10th IEEE International Conference on*, 2010, pp. 66–69.

[140] J. Müller, T. S. Böscke, S. Müller, E. Yurchuk, P. Polakowski, J. Paul, D. Martin, T. Schenk, K. Khullar, A. Kersch, W. Weinreich, S. Riedel, K. Seidel, A. Kumar, T. M. Arruda, S. V. Kalinin, T. Schloesser, R. Boschke, R. van Bentum, U. Schröder, and T. Mikolajick, "Ferroelectric Hafnium Oxide: A CMOS-compartible and highly scalable approach to future ferroelectric memories," in *Electron Devices Meeting (IEDM), IEEE International*, Washington, DC, USA, 2013, pp. 10.8.1 – 10.8.4.

[141] A. Kerber, E. Cartier, L. Pantisano, M. Rosmeulen, R. Degraeve, T. Kauerauf, G. Groeseneken, H. E. Maes, and U. Schwalke, "Characterization of the V_T-instability in SiO_2/HfO_2 gate dielectrics," in *Reliability Physics Symposium, 41rd Annual. 2003 IEEE International*, Dallas, Texas, 2003, pp. 41–45.

[142] G. Bersuker, J. Sim, C. S. Park, C. Young, S. Nadkarni, R. Choi, and B. H. Lee, "Mechanism of Electron Trapping and Characteristics of Traps in HfO_2 Gate Stacks," *Device and Materials Reliability, IEEE Transactions on*, vol. 7, no. 1, pp. 138–145, 2007.

[143] G. Bersuker, J. H. Sim, C. D. Young, R. Choi, P. M. Zeitzoff, G. A. Brown, B. H. Lee, and R. W. Murto, "Effect of Pre-Existing Defects on Reliability Assessment of high-*k* gate dielectrics," *Microelectronics Engineering*, vol. 44, pp. 1509–1512, 2004.

[144] J. Robertson, "High dielectric constant gate oxides for metal oxide Si transistors," *Reports on Progress in Physics*, vol. 69, no. 2, pp. 327–396, Feb. 2006.

[145] Y. P. Feng, A. T. L. Lim, and M. F. Li, "Negative-U property of oxygen vacancy in cubic HfO_2," *Applied Physics Letters*, vol. 87, no. 6, p. 062105, 2005.

[146] W.-T. Lu, P.-C. Lin, T.-Y. Huang, C.-H. Chien, M.-J. Yang, I.-J. Huang, and P. Lehnen, "The characteristics of hole trapping in HfO_2/SiO_2 gate dielectrics with TiN gate electrode," *Applied Physics Letters*, vol. 85, no. 16, pp. 3525–3527, 2004.

[147] G. Bersuker, J. Barnett, N. Moumen, B. Foran, C. D. Young, P. Lysaght, J. Peterson, B. H. Lee, P. Zeitzoff, and H. R. Huff, "Interfacial Layer-Induced Mobility Degradation in High-*k* Transistors," *Japanese Journal of Applied Physics*, vol. 43, no. 11B, pp. 7899–7902, 2004.

[148] C. D. Young, G. Bersuker, G. A. Brown, P. Lysaght, P. Zeitzoff, R. W. Murto, and H. R. Huff, "Charge trapping and device performance degradation in MOCVD hafnium-

based gate dielectric stack structures," in *Reliability Physics Symposium Proceedings, 42nd Annual. 2004 IEEE International*, 2004, pp. 597–598.

[149] C. D. Young, R. Choi, J. H. Sim, B. H. Lee, P. Zeitzoff, Y. Zhao, K. Matthews, G. A. Brown, and G. Bersuker, "Interfacial layer dependence of $HFSI_xO_y$ gate stacks on V_T instability and charge trapping using ultra-short pulse in characterization," in *Reliability Physics Symposium, 2005. Proceedings. 43rd Annual. 2005 IEEE International*, 2005, pp. 75–79.

[150] G. Ribes, J. Mitard, M. Denais, S. Bruyere, F. Monsieur, C. Parthasarathy, E. Vincent, and G. Ghibaudo, "Review on high-k dielectrics reliability issues," *IEEE Transactions on Device and Materials Reliability*, vol. 5, no. 1, pp. 5–19, Mar. 2005.

[151] R. Degraeve, M. Aoulaiche, B. Kaczer, P. Roussel, T. Kauerauf, S. Sahhaf, and G. Groeseneken, "Review of reliability issues in high-k/metal gate stacks," in *Physical and Failure Analysis of Integrated Circuits, 2008. IPFA 2008. 15th International Symposium on the*, 2008, pp. 1–6.

[152] K. Torii, H. Kitajima, T. Arikado, K. Shiraishi, S. Miyazaki, K. Yamabe, M. Boero, T. Chikyow, and K. Yamada, "Physical model of BTI, TDDB and SILC in HfO_2-based high-k gate dielectrics," in *Electron Devices Meeting, 2004. IEDM Technical Digest. IEEE International*, 2004, pp. 129–132.

[153] B. H. Lee, C. D. Young, R. Choi, J. H. Sim, G. Bersuker, C. Y. Kang, R. Harris, G. A. Brown, K. Matthews, and S. C. Song, "Intrinsic characteristics of high-k devices and implications of fast transient charging effects (FTCE)," in *Electron Devices Meeting, 2004. IEDM Technical Digest. IEEE International*, 2004, pp. 859–862.

[154] D. Heh, C. D. Young, and G. Bersuker, "Experimental Evidence of the Fast and Slow Charge Trapping/Detrapping Processes in High-k Dielectrics Subjected to PBTI Stress," *Electron Device Letters, IEEE*, vol. 29, pp. 180–182, 2008.

[155] H.-W. You and W.-J. Cho, "Charge trapping properties of the HfO_2 layer with various thicknesses for charge trap flash memory applications," *Applied Physics Letters*, vol. 96, no. 9, p. 093506, 2010.

[156] J. Buckley, M. Bocquet, G. Molas, M. Gely, P. Brianceau, N. Rochat, E. Martinez, F. Martin, H. Grampeix, and J. P. Colonna, "In-depth investigation of Hf-based high-k dielectrics as storage layer of charge-trap NVMs," in *Electron Devices Meeting, 2006. IEDM'06. International*, 2006, pp. 1–4.

[157] T. Sugizaki, M. Kobayashi, M. Ishidao, H. Minakata, M. Yamaguchi, Y. Tamura, Y. Sugiyama, T. Nakanishi, and H. Tanaka, "Novel multi-bit SONOS type flash memory using a high-k charge trapping layer," in *VLSI Technology, 2003. Digest of Technical Papers. 2003 Symposium on*, 2003, pp. 27–28.

[158] Yan Ny Tan, W. K. Chim, Wee Kiong Choi, Moon Sig Joo, and Byung Jin Cho, "Hafnium aluminum oxide as charge storage and blocking-oxide layers in SONOS-

type nonvolatile memory for high-speed operation," *IEEE Transactions on Electron Devices*, vol. 53, no. 4, pp. 654–662, Apr. 2006.

[159] J. L. Gavartin, D. Muñoz Ramo, A. L. Shluger, G. Bersuker, and B. H. Lee, "Negative oxygen vacancies in HfO_2 as charge traps in high-k stacks," *Applied Physics Letters*, vol. 89, no. 8, p. 082908, 2006.

[160] H. Park, M. Jo, H. Choi, M. Hasan, R. Choi, P. D. Kirsch, C. Y. Kang, B. H. Lee, T.-W. Kim, T. Lee, and H. Hwang, "The Effect of Nanoscale Nonuniformity of Oxygen Vacancy on Electrical and Reliability Characteristics of HfO_2 MOSFET Devices," *IEEE Electron Device Letters*, vol. 29, no. 1, pp. 54–56, Jan. 2008.

[161] N. A. Chowdhury and D. Misra, "Charge Trapping at Deep States in Hf–Silicate Based High-κ Gate Dielectrics," *Journal of The Electrochemical Society*, vol. 154, no. 2, p. G30, 2007.

[162] A. Foster, F. Lopez Gejo, A. Shluger, and R. Nieminen, "Vacancy and interstitial defects in hafnia," *Physical Review B*, vol. 65, no. 17, May 2002.

[163] H. Takeuchi, H. Y. Wong, D. Ha, and T.-J. King, "Impact of oxygen vacancies on high-κ gate stack engineering," in *Electron Devices Meeting, 2004. IEDM Technical Digest. IEEE International*, 2004, pp. 829–832.

[164] C. Shen, M. F. Li, X. P. Wang, H. Y. Yu, Y. P. Feng, A. T. L. Lim, Y. C. Yeo, D. S. H. Chan, and D. L. Kwong, "Negative U traps in HfO_2 gate dielectrics and frequency dependence of dynamic BTI in MOSFETs," in *Electron Devices Meeting, 2004. IEDM Technical Digest. IEEE International*, 2004, pp. 733–736.

[165] A. Ortiz-Conde, F. J. Garcıa Sánchez, J. J. Liou, A. Cerdeira, M. Estrada, and Y. Yue, "A review of recent MOSFET threshold voltage extraction methods," *Microelectronics Reliability*, vol. 42, no. 4, pp. 583–596, 2002.

[166] C. B. Sawyer and C. H. Tower, "Rochelle salt as dielectric," *Phys. Rev.*, vol. 35, p. 269, 1930.

[167] A. Gruverman and M. Alexe, *Nanoscale Characterization of Ferroelectric Materials*. Springer, 2004.

[168] *Manual aixACCT TF Analyzer 3000 with FE-Module*. aixACCT Systems GmbH, 2012.

[169] S. Bernacki, L. Jack, Y. Kisler, S. Collins, S. D. Bernstein, R. Hallock, B. Armstrong, J. Shaw, J. Evans, B. Tuttle, B. Hammetter, S. Rogers, B. Nasby, J. Henderson, J. Benedetto, R. Moore, C. R. Pugh, and A. Fennelly, "Standardized ferroelectric capacitor test methodology for nonvolatile semiconductor memory applications," *Integrated Ferroelectrics*, vol. 3, no. 2, pp. 97–112, Jun. 1993.

[170] P. K. Larsen, G. L. M. Kampschoer, M. B. van der Mark, and M. Klee, "Ultrafast polarization switching of lead zirconate titanate thin films," in *Applications of*

Ferroelectrics, 1992. ISAF'92., Proceedings of the Eighth IEEE International Symposium on, 1992, pp. 217–224.

[171] S. V. Kalinin, B. J. Rodriguez, S. Jesse, P. Maksymovych, K. Seal, M. Nikiforov, A. P. Baddorf, A. Kholkin, and R. Proksch, "Local bias-induced phase transition," *Materialstoday*, vol. 11, no. 11, pp. 16–27, 2008.

[172] P. Güthner and K. Dransfeld, "Local poling of ferroelectric polymers by scanning force microscopy," *Applied Physics Letters*, vol. 61, no. 9, p. 1137, 1992.

[173] A. Gruverman, O. Auciello, and H. Tokumoto, "Imaging and control of domain structures in ferroelectric thin films via scanning force microscopy," *Annual review of materials science*, vol. 28, no. 1, pp. 101–123, 1998.

[174] Y. Cho, S. Hashimoto, N. Odagawa, K. Tanaka, and Y. Hiranaga, "Nanodomain manipulation for ultrahigh density ferroelectric data storage," *Nanotechnology*, vol. 17, no. 7, pp. S137–S141, Apr. 2006.

[175] K. Tanaka, Y. Kurihashi, U. Tomoya, Y. Daimon, N. Odagawa, R. Hirose, Y. Hiranaga, and Y. Cho, "Scanning nonlinear dielectric microscopy nano-science and technology for next generation high density ferroelectric data storage," *Japanese Journal of Applied Physics*, vol. 47, no. 5, pp. 3311–3325, 2008.

[176] A. Gruverman, B. J. Rodriguez, C. Dehoff, J. D. Waldrep, A. I. Kingon, R. J. Nemanich, and J. S. Cross, "Direct studies of domain switching dynamics in thin film ferroelectric capacitors," *Applied Physics Letters*, vol. 87, no. 8, p. 082902, 2005.

[177] S. Jesse, A. P. Baddorf, and S. V. Kalinin, "Switching spectroscopy piezoresponse force microscopy of ferroelectric materials," *Applied Physics Letters*, vol. 88, no. 6, p. 062908, 2006.

[178] S. Jesse and S. V. Kalinin, "Band excitation in scanning probe microscopy: sines of change," *Journal of Physics D: Applied Physics*, vol. 44, no. 46, p. 464006, Nov. 2011.

[179] S. Jesse, P. Maksymovych, and S. V. Kalinin, "Rapid multidimensional data acquisition in scanning probe microscopy applied to local polarization dynamics and voltage dependent contact mechanics," *Applied Physics Letters*, vol. 93, no. 11, p. 112903, 2008.

[180] J. L. Autran, B. Balland, and G. Barbottin, "Charge pumping techniques: Their use for diagnosis and interface states studies in MOS transistors," *Instabilities in Silicon Devices*, vol. 3, pp. 405–493, 1999.

[181] J. S. Brugler and P. G. A. Jespers, "Charge Pumping in MOS Devices," *IEEE Trans. Electron. Dev.*, vol. ED-16, no. 3, pp. 297–302, 1969.

[182] G. Groeseneken, H. E. Maes, N. Beltran, and R. F. De Keersmaecker, "A reliable approach to charge-pumping measurements in MOS transistors," *Electron Devices, IEEE transactions on*, vol. 31, no. 1, pp. 42–53, 1984.

[183] G. Van den Bosch, G. Groeseneken, and H. E. Maes, "On the geometric component of charge-pumping current in MOSFETs," *Electron Device Letters, IEEE*, vol. 14, no. 3, pp. 107–109, 1993.

[184] A. Elliot, "The use of charge pumping currents to measure surface state densities in MOS transistors," *Solid-State Electronics*, vol. 19, no. 3, pp. 241–247, 1976.

[185] M. Declercq and P. Jespers, "Analysis of interface properties in MOS transistors by means of charge pumping measurments," *Revue HF*, vol. 9, no. 8, 1973.

[186] A. Kerber, E. Cartier, R. Degraeve, G. Groeseneken, H. E. Maes, and U. Schwalke, "Charge trapping in SiO_2/HfO_2 gate dielectrics: Comparison between charge-pumping and pulsed I_D–V_G," *Microelectronic Engineering*, vol. 72, pp. 267–272, 2004.

[187] C. Y. Lu, K. S. Chang-Liao, P. H. Tsai, and T. K. Wang, "Depth Profiling of Border Traps in MOSFET With High-*k* Gate Dielectric by Charge-Pumping Technique," *Electron Device Letters, IEEE*, vol. 27, no. 10, pp. 859–862, 2006.

[188] T. H. Hou, M. F. Wang, K. L. Mai, Y. M. Lin, M. H. Yang, L. G. Yao, Y. Jin, S. C. Chen, and M. S. Liang, "Direct Determination of Interface and Bulk Traps in Stacked HfO_2 Dielectrics Using Charge Pumping Method," *42th Annual International Reliability Physics Symposium*, pp. 581–582, 2004.

[189] D. Heh, C. D. Young, G. A. Brown, P. Y. Hung, A. Diebold, E. M. Vogel, J. B. Bernstein, and G. Bersuker, "Spatial Distributions of Trapping Centers in HfO_2/SiO_2 Gate Stack," *Electron Devices, IEEE Transactions on*, vol. 54, no. 6, pp. 1338–1345, 2007.

[190] C. D. Young, D. Heh, S. V. Nadkarni, R. Choi, J. J. Peterson, J. Barnett, B. H. Lee, and G. Bersuker, "Electron trap generation in high-*k* gate stacks by constant voltage stress," *Device and Materials Reliability, IEEE Transactions on*, vol. 6, no. 2, pp. 123–131, 2006.

[191] M. Masuduzzaman, A. E. Islam, and M. A. Alam, "Exploring the capability of multifrequency charge pumping in resolving location and energy levels of traps within dielectric," *Electron Devices, IEEE Transactions on*, vol. 55, no. 12, pp. 3421–3431, 2008.

[192] S. Sahhaf, R. Degraeve, M. Cho, K. D. Brabanter, P. J. Roussel, M. B. Zahid, and G. Groeseneken, "Detailed analysis of charge pumping and $I_d V_g$ hysteresis for profiling traps in SiO_2/HfSiO(N)," *Microelectronic Engineering*, vol. 87, no. 12, pp. 2614–2619, 2010.

[193] R. E. Paulsen and M. H. White, "Theory and application of charge pumping for the characterization of Si-SiO_2 interface and near-interface oxide traps," *Electron Devices, IEEE Transactions on*, vol. 41, no. 7, pp. 1213–1216, 1994.

[194] R. E. Paulsen, R. R. Siergiej, M. L. French, and M. H. White, "Observation of near-interface oxide traps with the charge-pumping technique," *Electron Device Letters, IEEE*, vol. 13, no. 12, pp. 627–629, 1992.

[195] C. D. Young, D. Heh, R. Choi, B. H. Lee, and G. Bersuker, "The Pulsed I_d-V_g methodology and Its Application to the Electron Trapping Characterization of High-κ gate Dielectrics," *Journal of Semiconductor Technology and Science*, vol. 10, no. 2, p. 79, 2010.

[196] C. Leroux, J. Mitard, G. Ghibaudo, X. Garros, G. Reimbold, B. Guillaumor, and F. Martin, "Characterization and modeling of hysteresis phenomena in high-*k* dielectrics," in *Electron Devices Meeting, 2004. IEDM Technical Digest. IEEE International*, 2004, pp. 737–740.

[197] C. D. Young, Y. Zhao, M. Pendley, B. H. Lee, K. Mathews, J. H. Sim, R. Choi, G. A. Brown, R. W. Murto, and G. Bersuker, "Ultra-Short Pulse Current–Voltage Characterization of the Intrinsic Characteristics of High-*k* Devices," *J.J.Appl. Phys.*, vol. 44, pp. 2437–2440, 2005.

[198] D. Heh, C. D. Young, R. Choi, and G. Bersuker, "Extraction of the threshold-voltage shift by the single-pulse technique," *Electron Device Letters, IEEE*, vol. 28, no. 8, pp. 734–736, 2007.

[199] R. E. Dinnebier and S. J. L. Billinge, *Powder Diffraction Theory and Practice*. Cambridge: The Royal Society of Chemistry, 2008.

[200] M. Birkholz, *Thin Film Analysis by X-Ray Scattering*. Weinheim: WILEY-VCH Verlag GmbH & Co. KGaA, 2006.

[201] L. Spieß, G. Teichert, R. Schwarzer, H. Behnken, and C. Genzel, *Moderne Röntgenbeugung*, 2. Auflage. Wiesbaden: Vieweg+Teubner, GWV Fachverlage GmbH, 2009.

[202] *HfO₂ tetragonal P42/nmc*, Powder Diffraction Card 04-011-8820. International Centre for Data Diffraction, 2011.

[203] *TiN cubic Fm3m*, Powder Diffraction Card 00-006-0642. International Centre for Data Diffraction, 2011.

[204] D. Brandon and W. D. Kaplan, *Microstructural Characterization of Materials*, 2nd Edition. John Wiley & Sons Ltd, 2008.

[205] D. B. Williams and C. B. Carter, "Part 3 Imaging," in *Transmission Electron Microscopy/ A Textbook for Materials Science*, Second Edition., Springer, 2009, pp. 371–578.

[206] B. Fultz and J. Howe, *Transmission Electron Microscopy and Diffractometry of Materials*, Fourth Edition. Berlin Heidelberg: Springer, 2013.

[207] J. Müller, "Ferroelektrizität in Hafniumdioxid und deren Anwendung in nicht-flüchtigen Halbleiterspeichern," Doctoral dissertation, Technische Universität Dresden, 2014.

[208] Y. Pu, J. Zhu, X. Zhu, Y. Luo, M. Wang, X. Li, J. Liu, J. Zhu, and D. Xiao, "Double hysteresis loop induced by defect dipoles in ferroelectric $Pb(Zr_{0.8}Ti_{0.2})O_3$ thin films," *Journal of Applied Physics*, vol. 109, no. 4, p. 044102, 2011.

[209] K. Carl and K. H. Hardtl, "Electrical after-effects in $Pb(Ti,Zr)O_3$ ceramics," *Ferroelectrics*, vol. 17, no. 1, pp. 473–486, Jan. 1977.

[210] W. J. Merz, "Double Hysteresis Loop of $BaTiO_3$ at the Curie Point," *Physical Review*, vol. 91, no. 3, p. 513, 1953.

[211] C. Fachmann, L. Frey, S. Kudelka, T. Boescke, S. Nawka, E. Erben, and T. Doll, "Tuning the dielectric properties of hafnium silicate films," *Microelectronic Engineering*, vol. 84, no. 12, pp. 2883–2887, Dec. 2007.

[212] A. I. Kingon, J.-P. Maria, and S. K. Streiffer, "Alternative dielectrics to silicon dioxide for memory and logic devices," *Nature*, vol. 406, no. 6799, pp. 1032–1038, 2000.

[213] A. Toriumi, Y. Yamamoto, Y. Zhao, K. Tomida, and K. Kita, "Doped HfO_2 for Higher-*k* Dielectrics," in *Proceedings of the 208th Elecrochemical Society (ECS) Meeting*, Los Angeles, California, 2005.

[214] W. J. Zhu, T. Tamagawa, M. Gibson, T. Furukawa, and T. P. Ma, "Effect of Al inclusion in HfO_2 on the physical and electrical properties of the dielectrics," *IEEE Electron Device Letters*, vol. 23, no. 11, pp. 649–651, Nov. 2002.

[215] *ZrO_2 orthorhombic Pbc21*, Powder Diffraction Card 04-005-4478. International Centre for Data Diffraction, 2011.

[216] *HfO_2 monoclinic P21/c*, Powder Diffraction Card 00-006-0318. International Centre for Data Diffraction, 2011.

[217] T. S. Böscke, S. Govindarajan, P. D. Kirsch, P. Y. Hung, C. Krug, B. H. Lee, J. Heitmann, U. Schröder, G. Pant, B. E. Gnade, and W. H. Krautschneider, "Stabilization of higher-κ tetragonal HfO_2 by SiO_2 admixture enabling thermally stable metal-insulator-metal capacitors," *Applied Physics Letters*, vol. 91, no. 7, p. 072902, 2007.

[218] D. Fischer and A. Kersch, "Stabilization of the high-*k* tetragonal phase in HfO_2: The influence of dopants and temperature from ab initio simulations," *Journal of Applied Physics*, vol. 104, no. 8, p. 084104, 2008.

[219] F. Bohra, B. Jiang, and J.-M. Zuo, "Textured crystallization of ultrathin hafnium oxide films on silicon substrate," *Applied Physics Letters*, vol. 90, no. 16, p. 161917, 2007.

[220] J. E. Lowther, J. K. Dewhurst, J. M. Leger, and J. Haines, "Relative stability of ZrO_2 and HfO_2 structural phases," *Physical Review B*, vol. 60, no. 21, p. 14485, 1999.

[221] T. D. Huan, V. Sharma, G. A. J. Rossetti, and R. Ramprasad, "Pathways Towards Ferroelectricity in Hafnia," *arXiv:1407.1008 (unpublished)*, 2014.

[222] S. E. Reyes-Lillo, K. F. Garrity, and K. M. Rabe, "Antiferroelectricity in thin film ZrO_2 from first principles," *arXiv:1403.3878v3 (unpublished)*, 2014.

[223] D. Martin, J. Müller, T. Schenk, T. M. Arruda, A. Kumar, E. Strelcov, E. Yurchuk, S. Müller, D. Pohl, A. Kersch, U. Schröder, S. V. Kalinin, and T. Mikolajick, "Ferroelectricity in Si-doped HfO_2 revealed: A binary lead-free ferroelectric," *to be published*.

[224] U. Schröder, E. Yurchuk, J. Müller, D. Martin, T. Schenk, P. Polakowski, C. Adelmann, M. Popovici, S. V. Kalinin, and T. Mikolajick, "Impact of Different Dopants on the Switching Properties of Ferroelectric Hafniumoxide," *Japanese Journal of Applied Physics*, vol. 53, p. 08LE02, 2014.

[225] H. Ikeda, T. Goto, M. Sakashita, A. Sakai, S. Zaima, and Y. Yasuda, "Local Leakage Current of HfO_2 Thin Films Characterized by Conducting Atomic Force Microscopy," *Japanese Journal of Applied Physics*, no. 42, pp. 1949–1953, 2003.

[226] E. P. Gusev, C. Cabral Jr, M. Copel, C. D'Emic, and M. Gribelyuk, "Ultrathin HfO_2 films grown on silicon by atomic layer deposition for advanced gate dielectrics applications," *Microelectronic Engineering*, vol. 69, no. 2, pp. 145–151, 2003.

[227] S. Toyoda, H. Takahashi, H. Kumigashira, M. Oshima, D.-I. Lee, S. Sun, Z. Liu, Y. Sun, P. A. Pianetta, I. Oshiyama, K. Tai, and S. Fukuda, "Study on mechanism of crystallization in HfO_2 films on Si substrates by in-depth profile analysis using photoemission spectroscopy," *Journal of Applied Physics*, vol. 106, no. 6, p. 064103, 2009.

[228] C. Richter, T. Schenk, U. Schroeder, and T. Mikolajick, "Influence of the ALD growth parameters on the ferroelectric properties of Si doped HfO_2," presented at the 12th International Baltic Conference on Atomic Layer Deposition, Helsinki, Finland, 2014.

[229] T. S. Böscke, J. Müller, D. Bräuhaus, U. Schröder, and U. Böttger, "Ferroelectricity in hafnium oxide: CMOS compatible ferroelectric field effect transistors," in *Electron Devices Meeting (IEDM), 2011 IEEE International*, 2011, pp. 24–5.

[230] D. H. Triyoso, R. I. Hegde, B. E. White, and P. J. Tobin, "Physical and electrical characteristics of atomic-layer-deposited hafnium dioxide formed using hafnium tetrachloride and tetrakis(ethylmethylaminohafnium)," *Journal of Applied Physics*, vol. 97, no. 12, p. 124107, 2005.

[231] T. Kawahara, K. Torii, R. Mitsuhashi, A. Muto, A. Horiuchi, H. Ito, and H. Kitajima, "Effect of Hf sources, oxidizing agents, and NH3/Ar plasma on the properties of HfAlOx films prepared by atomic layer deposition," *Japanese journal of applied physics*, vol. 43, no. 7R, p. 4129, 2004.

[232] J. Glaum, Y. A. Genenko, H. Kungl, L. Ana Schmitt, and T. Granzow, "De-aging of Fe-doped lead-zirconate-titanate ceramics by electric field cycling: 180°- vs. non-180° domain wall processes," *Journal of Applied Physics*, vol. 112, no. 3, p. 034103, 2012.

[233] S. K. Pandey, O. P. Thakur, A. Kumar, C. Prakash, R. Chatterjee, and T. C. Goel, "Study of pinched loop characteristics of lead zirconate titanate (65/35)," *Journal of Applied Physics*, vol. 100, no. 1, p. 014104, 2006.

[234] U. Robels and G. Arlt, "Domain wall clamping in ferroelectrics by orientation of defects," *Journal of Applied Physics*, vol. 73, no. 7, p. 3454, 1993.

[235] T. Schenk, M. Pesic, U. Schröder, M. Popovici, C. Adelmann, S. Van Elshocht, Y. V. Pershin, and T. Mikolajick, *to be published*.

[236] Y.-H. Kim and J. C. Lee, "Reliability characteristics of high-*k* dielectrics," *Microelectronics Reliability*, vol. 44, no. 2, pp. 183–193, Feb. 2004.

[237] K. T. Lee, J. Nam, M. Jin, K. Bae, J. Park, L. Hwang, J. Kim, H. Kim, and J. Park, "Frequency Dependent TDDB Behaviors and Its Reliability Qualification in 32nm High-*k*/Metal Gate CMOSFETs," presented at the Reliability Physics Symposium (IRPS), 2011 IEEE International, Monterey, CA, 2011, pp. 2A.3.1 – 2A.3.5.

[238] M. Masuduzzaman and M. A. Alam, "Hot Atom Damage (HAD) Limited TDDB Lifetime of Ferroelectric Memories," in *Electron Devices Meeting (IEDM), 2013 IEEE International*, 2013, pp. 21–24.

[239] B. P. Linder, S. Lombardo, J. H. Stathis, A. Vayshenker, and D. J. Frank, "Voltage dependence of hard breakdown growth and the reliability implication in thin dielectrics," *IEEE Electron Device Letters*, vol. 23, no. 11, pp. 661–663, Nov. 2002.

[240] D. P. Vijay and S. B. Desu, "Electrodes for $PbZr_xTi_{1-x}O_3$ Ferroelectric Thin Films," *Journal of the Electrochemical Society*, vol. 140, no. 9, pp. 2640–2645, 1993.

[241] C. A-Paz de Araujo, J. D. Cuchiaro, L. D. McMillan, M. C. Scott, and J. F. Scott, "Fatigue-free ferroelectric capacitors with platinum electrodes," *Nature*, vol. 374, pp. 627 – 629, 1995.

[242] C. Vallée, C. Mannequin, P. Gonon, L. Latu-Romain, H. Grampeix, and V. Jousseaume, "Resistive switching in metal-oxide-metal devices: fundamental understanding in relation to material characterization," in *Bipolar/BiCMOS Circuits and Technology Meeting (BCTM), 2013 IEEE*, 2013, pp. 151–158.

[243] T. Nakamura, Y. Nakao, A. Kamisawa, and H. Takasu, "Preparation of $Pb(Zr,Ti)O_3$ thin films on electrodes including IrO_2," *Applied Physics Letters*, vol. 65, no. 12, p. 1522, 1994.

[244] J. S. Cross, M. Fujiki, M. Tsukada, K. Matsuura, S. Otani, M. Tomotani, Y. Kataoka, Y. Kotaka, and Y. Goto, "Evaluation of PZT capacitors with $Pt/SrRuO_3$ electrodes for feram," *Integrated Ferroelectrics*, vol. 25, no. 1–4, pp. 265–273, Sep. 1999.

[245] J. Müller, T. S. Böscke, U. Schröder, R. Hoffmann, T. Mikolajick, and L. Frey, "Nanosecond Polarization Switching and Long Retention in a Novel MFIS-FET Based on Ferroelectric HfO_2," *IEEE Electron Device Letters*, vol. 33, no. 2, pp. 185–187, Feb. 2012.

[246] Y. Fujisaki, K. Iseki, and H. Ishiwara, "Long Retention Performance of a MFIS Device Achieved by Introducing High-k $Al_2O_3/Si_3N_4/Si$ Buffer Layer," *Material Reseach Society Symposium Proceedings*, vol. 786, p. E9.6/C9.6., 2004.

[247] S. Mueller, S. R. Summerfelt, J. Müller, U. Schroeder, and T. Mikolajick, "Ten-Nanometer Ferroelectric $Si:HfO_2$ Films for Next-Generation FRAM Capacitors," *IEEE Electron Device Letters*, vol. 33, no. 9, pp. 1300–1302, Sep. 2012.

[248] C. T. Black, C. Farrell, and T. J. Licata, "Suppression of ferroelectric polarization by an adjustable depolarization field," *Applied Physics Letters*, vol. 71, no. 14, p. 2041, 1997.

[249] A. Fayrushin, K. Seol, J. Na, S. Hur, J. Choi, and K. Kim, "The new program/erase cycling degradation mechanism of NAND Flash memory devices," in *Electron Devices Meeting (IEDM), IEEE International*, 2009, pp. 1–4.

[250] E. Yurchuk, J. Müller, J. Paul, R. Hoffmann, T. Schlösser, S. Müller, D. Martin, S. Slesazeck, R. Boschke, J. Sundqvist, R. van Bentum, M. Trentzsch, U. Schröder, and T. Mikolajick, "Origin of Endurance Degradation in the Novel HfO_2-based 1T Ferroelectric Non-Volatile Memories," in *Reliability Physics Symposium (IRPS), 2014 IEEE International*, Hawaii, USA, 2014, p. to be published.

[251] M. B. Zahid, R. Degraeve, L. Pantisano, J. F. Zhang, and G. Groeseneken, "Defects generation in SiO_2/HfO_2 studied with variable tcharge-tdischarge charge pumping (VT^2CP)," in *Reliability physics symposium, 2007. proceedings. 45th annual. ieee international*, 2007, pp. 55–60.

[252] T. Kauerauf, R. Degraeve, E. Cartier, B. Govoreanu, P. Blomme, B. Kaczer, L. Pantisano, A. Kerber, and G. Groeseneken, "Towards understanding degradation and breakdown of SiO_2/high-k stacks," presented at the Electron Devices Meeting, 2002. IEDM '02. International, San Francisco, CA, USA, 2002, pp. 521 – 524.

[253] E. Y. Wu, D. P. Ioannou, and C. B. LaRow, "Influence of Charge Trapping on Failure Detection and Its Distributions for nFET High-κ Stacks," presented at the Electron Devices Meeting (IEDM), 2011 IEEE International, Washington, DC, 2011, pp. 18.2.1 – 18.2.4.

[254] A. Kerber, A. Vayshenker, D. Lipp, T. Nigan, and E. Cartier, "Impact of charge trapping on the voltage acceleration of TDDB in metal gate/high-k n-channel MOSFETs," presented at the Reliability Physics Symposium (IRPS), 2010 IEEE International, Anaheim, CA, 2010, pp. 369 – 372.

[255] S. Knebel, S. Kupke, U. Schroeder, S. Slesazeck, T. Mikolajick, R. Agaiby, and M. Trentzsch, "Influence of Frequency Dependent Time to Breakdown on High-K/Metal Gate Reliability," *IEEE Transactions on Electron Devices*, vol. 60, no. 7, pp. 2368–2371, Jul. 2013.

[256] T. Hase, T. Noguchi, K. Takemura, and Y. Miyasaka, "Imprint Characteristics of $SrBi_2Ta_2O_9$ Thin Films with Modified Sr Composition," *Japanese Journal of Applied Physics*, vol. 37, pp. 5198–5202, 1998.

[257] R. G. Southwick, A. Sup, A. Jain, and W. B. Knowlton, "An Interactive Simulation Tool for Complex Multilayer Dielectric Devices," *IEEE Transactions on Device and Materials Reliability*, vol. 11, no. 2, pp. 236–243, Jun. 2011.

[258] R. G. Soutwick, A. Sup, A. Jain, W. B. Knowlton, and M. Baker, *Multi-Dielectric Energy Band Diagram Program.* (available online at http://nano.boisestate.edu/research-areas/multi-dielectric-energy-band-diagram-program/): Knowlton Research Group -Boise State University, 2014.

[259] L. D. Yau, "A simple theory to predict the threshold voltage of short-channel IGFET's," *Solid-State Electronics*, vol. 17, no. 10, pp. 1059–1063, 1974.

[260] S. Inaba, K. Okano, S. Matsuda, M. Fujiwara, A. Hokazono, K. Adachi, K. Ohuchi, H. Suto, H. Fukui, T. Shimizu, S. Mori, H. Oguma, A. Murakoshi, T. Itani, T. Iinuma, T. Kudo, H. Shibata, S. Taniguchi, M. Takayanagi, A. Azuma, H. Oyamatsu, K. Suguro, Y. Katsumata, Y. Toyoshima, and H. Ishiuchi, "High performance 35 nm gate length CMOS with NO oxynitride gate dielectric and Ni salicide," *IEEE Transactions on Electron Devices*, vol. 49, no. 12, pp. 2263–2270, Dec. 2002.

[261] T. Ito, K. Suguro, T. Itani, K. Nishinohara, K. Matsuo, and T. Saito, "Improvement of threshold voltage roll-off by ultra-shallow junction formed by flash lamp annealing," in *VLSI Technology, 2003. Digest of Technical Papers. 2003 Symposium on*, 2003, pp. 53–54.

[262] T. Hatanaka, M. Takahashi, S. Sakai, and K. Takeuchi, "A zero V_{TH} memory cell ferroelectric-NAND flash memory with 32% read disturb, 24% program disturb, 10% data retention improvement for enterprise SSD," in *Solid State Device Research Conference, ESSDERC'09. Proceedings of the European*, 2009, pp. 225–228.

[263] A. Gruverman, "Scaling effect on statistical behavior of switching parameters of ferroelectric capacitors," *Applied Physics Letters*, vol. 75, no. 10, p. 1452, 1999.

[264] D. Heh, C. D. Young, and G. Bersuker, "Experimental Evidence of the Fast and Slow Charge Trapping/Detrapping Processes in High-k Dielectrics Subjected to PBTI Stress," *Electron Device Letters, IEEE*, vol. 29, no. 2, pp. 180–182, 2008.

[265] D. Heh, R. Choi, C. D. Young, and G. Bersuker, "Fast and slow charge trapping/detrapping processes in high-k nMOSFETs," in *Integrated Reliability Workshop Final Report, 2006 IEEE International*, 2006, pp. 120–124.

[266] R. Choi, S. J. Rhee, J. C. Lee, B. H. Lee, and G. Bersuker, "Charge trapping and detrapping characteristics in hafnium silicate gate stack under static and dynamic stress," *Electron Device Letters, IEEE*, vol. 26, no. 3, pp. 197–199, 2005.

[267] R. Choi, S. C. Song, C. D. Young, G. Bersuker, and B. H. Lee, "Charge trapping and detrapping characteristics in hafnium silicate gate dielectric using an inversion pulse measurement technique," *Applied Physics Letters*, vol. 87, no. 12, pp. 122901– 122901, 2005.

Bibliography

List of symbols and abbreviations

Symbol	Unit	Description
A	m^2	Area
A_{eff}	m^2	Effective area of the channel
A_F	m^2	Area of a ferroelectric capacitor
C	$F\,cm^{-2}$	Capacitance per area
C_F	$F\,cm^{-2}$	Ferroelectric capacitance per area
C_i	$F\,cm^{-2}$	Insulator capacitance per area
C_P	$F\,cm^{-2}$	Parallel capacitance per area
C_R	$F\,cm^{-2}$	Capacitance of a reference capacitor per area
C_S	$F\,cm^{-2}$	Series capacitance per area
D	–	Dissipation factor
d	nm	Film thickness
d_{FE}	nm	Thickness of a ferroelectric layer
d_{hkl}	nm	Spacing between the series of parallel lattice planes with Miller indices (h, k, l)
d_{IL}	nm	Thickness of an insulating layer
D_{it}	$eV^{-1}cm^{-2}$	Mean energetic density of surface states
E	$MV\,cm^{-1}$	Electric field strength
E_C	$MV\,cm^{-1}$	Coercive electric field strength
E_{CR}	$MV\,cm^{-1}$	Critical Field
E_{DEP}	$MV\,cm^{-1}$	Depolarisation field
E_{EXT}	$MV\,cm^{-1}$	Field strength of external electric
E_g	eV	Bandgap
E_T	eV	Trap energy level
E_α	eV	Switching activation field
f	Hz	Frequency
I	A	Current
I_{CP}	A	Charge pumping current
I_{CP_MAX}	A	Maximum charge pumping current
I_D	A	Drain current
I_G	A	Gate current
I_{TH}	A	Drain current at threshold voltage

j	$A\ cm^{-1}$	Current density
j_G	$A\ cm^{-1}$	Gate current density
k	$eV\ K^{-1}$	Boltzmann constant
L	m	Transistor channel length
L_G	m	Transistor gate length
MW	V	Memory window
N_A	cm^{-3}	Acceptor impurity concentration
N_{CP}	cm^{-2}	Average surface density of interface states
n_i	cm^{-3}	Intrinsic carrier concentration of a semiconductor
P	$C\ cm^{-2}$	Polarisation per area
P_R	$C\ cm^{-2}$	Remanent polarisation per area
P_R-	$C\ cm^{-2}$	Negative remanent polarisation per area
P_R+	$C\ cm^{-2}$	Positive remanent polarisation per area
P_{R_rel}	$C\ cm^{-2}$	Relaxed remanent polarisation per area after 1 s
$PRampl$	nm	Piezoresponse amplitude
$PRphase$	$degree, °$	Piezoresponse amplitude
P_S	$C\ cm^{-2}$	Spontaneous polarisation
Q	$C\ cm^{-2}$	Charge per area
q	C	Unit electric charge
Q_{eff}	$C\ cm^{-2}$	Effective surface charge density per area
R	Ω	Resistance
R_{LOAD}	Ω	Load resistance
R_P	Ω	Parallel resistance
R_S	Ω	Series resistance
T	$°C$	Temperature
t	s	Time
t_{accum}	s	Time in accumulation
T_C	K	Curie temperature
t_{DTP}	s	Detrapping pulse width
$t_{DTP\ 100\%}$	s	Time required for a complete detrapping
t_{ERASE}	s	Erase pulse width
t_{ERASE_eff}	s	Effective erase time
t_f	s	Fall time
t_{FG}		Time of forward domain growth
t_{inv}	s	Time in inversion
t_r	s	Rise time
t_{SG}	s	Time of sideways domain growth

t_{SWITCH}	s	Time of ferroelectric switching
t_{TP}	s	Trapping pulse width
$t_{TP\,0}$	s	Trapping onset time
t_{WRITE}	s	Width of a writing pulse (program or erase)
V	V	Voltage
v	$m\ s^{-1}$	Speed of sound
V_C	V	Coercive voltage
V_{C-}	V	Negative coercive voltage
V_{C+}	V	Positive coercive voltage
V_D	V	Drain voltage
V_{DTP}	V	Voltage of a detrapping pulse
V_{ERASE}	V	Erase voltage
V_{FB}	V	Flatband voltage
V_G	V	Gate voltage
V_{GH}	V	High gate voltage level
V_{GL}	V	Low gate voltage level
V_R	V	Voltage drop over a reference capacitor
V_{TH}	V	Threshold voltage
$v_{th,}$	$m\ s^{-1}$	Thermal velocity of the carriers
V_{TH_PR-}	V	Threshold voltage of a ferroelectric transistor in a negatively polarised state
V_{TH_PR+}	V	Threshold voltage of a ferroelectric transistor in a positively polarised state
V_{TP}	V	Voltage of a trapping pulse
V_{WRITE}	V	Writing (program or erase) voltage
W	m	Transistor channel width
W_G	m	Transistor gate width
X	Ω	Reactance
θ	$degree,\ °$	Diffraction angle
λ	nm	Wavelength
μ_n	$cm^2\ V^{-1}s^{-1}$	Electron mobility
$\tilde{\sigma}$	cm^{-2}	Geometric mean value of the capture cross section
σ_n	cm^{-2}	Capture cross sections for electrons
σ_p	cm^{-2}	Capture cross sections for holes
τ	$-$	Reduced temperature $(T_C\text{-}T)/T_C$
ω	$degree,\ °$	Incidence angle
ΔP_R	$C\ cm^{-2}$	Loss of the remanent polarisation within 1 s

ΔV_G	V	Gate pulse amplitude
ΔV_{TH}	V	Threshold voltage shift
Φ_M	eV	Metal work function
Φ_{MS}	eV	Metal-semiconductor work function difference
χ^e	eV	Electron affinity
ε_0	$F\ cm^{-1}$	Permittivity of vacuum
ε_{FE}	$-$	Relative permittivity of a ferroelectric
ε_{IL}	$-$	Relative permittivity of an insulator
ε_S	$-$	Relative semiconductor permittivity
ψ_B	V	Fermi potential
ψ_S	V	Semiconductor surface potential

Abbreviation   Description

AC	Alternating Current
AFE	Antiferroelectric
Al	Aluminium
CMOS	Complementary Metal-Oxide-Silicon
CP	Charge Pumping
CT	Charge-Trapping
CVD	Chemical Vapour Deposition
DC	Direct Current
DRAM	Dynamic Random Access Memory
EEPROM	Electrically Erasable and Programmable Read Only Memory
EPROM	Electrically Programmable Read Only Memory
FE	Ferroelectric
FeFET	Ferroelectric Field Effect Transistor
FeRAM	Ferroelectric Random Access Memory
FG	Floating Gate
FN	Fowler-Nordheim
FWHM	Full Width at Half Maximum
GCR	Gate Coupling Factor
Gd	Gadolinium
GI-XRD	Grazing Incidence X-Ray Diffraction
H_2O	Water
$HfCl_2$	Hafnium Tetrachloride

HfO$_2$	Hafnium oxide
HfSiON	Hafnium Silicon Oxynitride
ICDD	International Centre for Diffraction Data
IrO$_2$	Iridium Oxide
ITRS	International Technology Roadmap of Semiconductors
La	Lanthanum
m	Monoclinic Phase
MFIS	Metal-Ferroelectric-Insulator-Semiconductor
MFIS-FET	Metal-Ferroelectric-Insulator-Semiconductor Field Effect Transistor
MFM	Metal-Ferroelectric-Metal
MFMIS	Metal-Ferroelectric-Metal-Insulator
MIM	Metal-Insulator-Metal
MIS	Metal-Insulator-Semiconductor
MISFET	Metal-Insulator-Semiconductor Field Effect Transistor
MOSFET	Metal-Oxide-Semiconductor Field Effect Transistor
MRAM	Magnetic Random Access Memory
N$_2$	Nitrogen Gas
NH$_3$	Ammonia Gas
NLS	Nucleation-Limited-Switching
NVM	Non-Volatile Memory
NVSM	Non-Volatile Semiconductor Memory
o	Orthorhombic Phase
O$_3$	Ozone
PCRAM	Phase Change Random Access Memory
PFM	Piezoresponse Force Microscopy
PMU	Pulse Measurement Unit
Poly-Si	Polycrystalline Silicon
Pt	Platinum
PUND	Positive Up Negative Down
PVD	Physical Vapour Deposition
PZT	Lead Zirconate Titanate
RAM	Random Access Memory
ReRAM	Resistive Random Access Memory
ROM	Read Only Memory
RPM	Remote Pulse Amplifier
RuO$_2$	Ruthenium Oxide
SBT	Strontium Bismuth Tantalate

List of symbols and abbreviations

Si	Silicon
Si:HfO$_2$	Silicon doped Hafnium Oxide
SiCl$_4$	Silicon Tetrachloride
SiN	Silicon Nitride
SiON	Silicon Oxynitride
SONOS	Silicon-Oxide-Nitride-Oxide-Semiconductor
Sr	Strontium
SRAM	Static Random Access Memory
t	Tetragonal Phase
TEM	Transmission Electron Spectroscopy
TiCl$_4$	Titanium Tetracloride
TiN	Titanium Nitride
XPS	X-Ray Photoelectron Spectroscopy
XRD	X-Ray Diffraction
Y	Yttrium
Zr	Zirconium
ZrO$_2$	Zirconium Oxide

List of Publications

Journal Articles:

Characterisation of retention properties of charge-trapping memory cells at low temperatures
E. Yurchuk, J. Bollmann and T. Mikolajick, Materials Science and Engineering **5** 012026 (2009)

Impact of layer thickness on the ferroelectric behaviour of silicon doped hafnium oxide thin films
E. Yurchuk, J. Müller, S. Knebel, J. Sundqvist, A. P. Graham, T. Melde, U. Schröder, T. Mikolajick, Thin Solid Films **533** 88 (2013)

Downscaling ferroelectric field effect transistors by using ferroelectric Si-doped HfO₂
D. Martin, E. Yurchuk, S. Müller, J. Müller, J. Paul, J. Sundquist, S. Slesazeck, T. Schlösser, R. van Bentum, M. Trentzsch, U. Schröder, T. Mikolajick, Solid-State Electronics **88** 65 (2013)

From MFM Capacitors Toward Ferroelectric Transistors: Endurance and Disturb Characteristics of HfO₂-Based FeFET Devices
S. Mueller, J. Muller, R. Hoffmann, E. Yurchuk, T. Schlosser, R. Boschke; J. Paul, M. Goldbach; T. Herrmann, A. Zaka; U. Schroder and T.Mikolajick, Electron Devices, IEEE Transactions on **60** 4199 (2013)

Hafnium Oxide Based CMOS Compatible Ferroelectric Materials
U. Schroeder, S. Mueller, J. Mueller, E. Yurchuk, D. Martin, C. Adelmann, T. Schloesser, R. van Bentum and T. Mikolajick, ECS Journal of Solid State Science and Technology **2** N69 (2013)

Impact of Different Dopants on the Switching Properties of Ferroelectric Hafniumoxide
U. Schroeder, E. Yurchuk, J. Müller, D. Martin, T. Schenk, C. Adelmann, M. Popovici, S. Kalinin, T. Mikolajick, Japanese Journal of Applied Physics **53** 08LE02 (2014)

163

Conference Contributions:

Characterisation of retention properties of charge-trapping memory cells at low temperatures
E. Yurchuk, J. Bollmann and T. Mikolajick, 5th International EEIGM/AMASE/FORGEMAT
Conference on Advanced Materials Research, Nancy, France (2009)

An empirical model describing the MLC retention of charge trap flash memories
T. Melde, R. Hoffmann, E. Yurchuk, J. Paul, and T. Mikolajick, IEEE International Integrated
Reliability Workshop (IRW), Stanford Sierra, CA USA (2010)

Evaluation of measurement techniques for characterization of charge trapping materials for memory applications
E. Yurchuk, T. Melde and T. Mikolajick, DPG Spring Meeting, Dresden, Germany (2011)

Influence of silicon content, annealing temperature and film thickness on the emergence of ferroelectricity in HfO_2 and its implication on highly scaled Ferroelectric Field Effect Transistors
E. Yurchuk, J. Müller, S. Knebel, R. Hoffmann, T. Melde, S. Müller, D. Martin, S. Slesazeck,
J. Sundqvist, R. Boschke, T. Schlösser, R. van Bentum, M. Trentzsch, U. Schröder and
T. Mikolajick, European Materials Research Society (E-MRS) Spring Meeting, Strasbourg,
France (2012)

HfO_2-based Ferroelectric Field-Effect Transistors with 260 nm channel length and long data retention
E. Yurchuk, J. Müller, R. Hoffmann, J. Paul, D. Martin, R. Boschke, T. Schlösser,
S. Müller, S. Slesazeck, R. van Bentum, M. Trentzsch, U. Schröder and T. Mikolajick,
4th IEEE International Memory Workshop (IMW), Milan, Italy (2012)

Ferroelectricity in HfO_2 enables nonvolatile data storage in 28 nm HKMG
J. Müller, E. Yurchuk ,T. Schlösser, J. Paul, R. Hoffmann, S. Müller, D. Martin, S. Slesazeck,
P. Polakowski, J. Sundqvist, M. Czernohorsky, K. Seidel, P. Kücher, R. Boschke,
M.Trentzsch, K. Gebauer, U. Schröder, T. Mikolajick, Symposium on VLSI Technology,
Honolulu, HI USA (2012)

Downscaling Ferroelectric Field Effect Transistors by using ferroelectric Si-doped HfO_2
D. Martin, E. Yurchuk, S. Muller, J. Muller, J. Paul, J. Sundquist, S. Slesazeck, T. Schloesser,
R. van Bentum, M. Trentzsch, U. Schroeder and T. Mikojajick, 13th IEEE Conference on
Ultimate Integration on Silicon (ULIS), Grenoble, France (2012)

Charge trapping in Si:HfO₂-based ferroelectric field effect transistors: A fast transient characterization using pulsed Id-Vg methodology

J. Müller, E. Yurchuk, T. S. Böscke, R. Hoffmann, U. Schröder,T. Mikolajick, L. Frey, 17th Workshop on Dielectrics in Microelectronics(WoDiM), Dresden, Germany (2012)

Applicability of Ferroelectric HfO₂ for Non-Volatile Memory-Cell Arrays

S. Müller, E. Yurchuk, J. Müller, S. Slesazeck, T. Schlösser, D. Martin, R. Hoffmann, J. Paul, R. Boschke, R. van Bentum, M. Trentzsch, U. Schröder and T. Mikolajick, 17th Workshop on Dielectrics in Microelectronics (WoDiM), Dresden, Germany (2012)

Non-volatile data storage in HfO₂-based ferroelectric FETs

U Schroeder, E Yurchuk, S Mueller, J Mueller, S Slesazeck, T Schloesser, M Trentzsch, T Mikolajick, 12th Annual Non-Volatile Memory Technology Symposium (NVMTS), Singapore (2012)

Ferroelectric Hafnium Oxide: A CMOS-compatible and highly scalable approach to future ferroelectric memories

J. Müller, T.S. Böscke, S. Müller, E. Yurchuk, P. Polakowski, J. Paul, D. Martin, T. Schenk, K. Khullar, A. Kersch, W. Weinreich, S. Riedel, K. Seidel, A. Kumar, T.M. Arruda, S.V. Kalinin, T. Schlösser, R. Boschke, R. van Bentum, U. Schröder, T. Mikolajick, IEEE International Electron Devices Meeting (IEDM), Washington, DC USA (2013)

Performance investigation and optimization of Si:HfO₂ FeFETs on a 28 nm bulk technology

S. Mueller, E. Yurchuk, S. Slesazeck, T. Mikolajick, J. Muller, T. Herrmann, A. Zaka, IEEE International Symposium on the Applications of Ferroelectric and Workshop on the Piezoresponse Force Microscopy (ISAF/PFM), Prague, Czech Republic (2013)

Origin of Endurance Degradation in the Novel HfO₂-based 1T Ferroelectric Non-Volatile Memories

E. Yurchuk, J. Müller, J. Paul, R. Hoffmann, S. Müller, D. Martin, S. Slesazeck, U. Schröder, J. Sundqvist, T. Schlösser, R. Boschke, R. van Bentum, M. Trentzsch, T. Mikolajick, IEEE International Reliability Physics Symposium (IRPS), Waikoloa Village, HI USA (2014)

Influence of Charge Trapping on Memory Characteristics of Si:HfO₂-Based Ferroelectric Field Effect Transistors

M. Pešić, S. Mueller, S. Slesazeck, A. Zaka, T. Herrmann, E. Yurchuk, U. Schröder, T. Mikolajick, 4rd International Workshop on Simulation and Modeling of Memory Devices (IWSMM), Milan, Italy, (2013)

Ferroelectricity in Doped Hafnium Oxide
U. Schroeder, E. Yurchuk, J. Müller, D. Martin, T. Schenk, C. Adelmann, S. Kalinin, U. Boettger, A. Kersch, T. Mikolajick, IEEE International Symposium on the Applications of Ferroelectric (ISAF), State College PA, USA (2014)

Doped Hafnium Oxide –An Enabler for Ferroelectric Field Effect Transistors
T. Mikolajick, S. Müller, T. Schenk, E. Yurchuk, S. Slesazeck, U. Schröder, S. Flachowsky, R. van Bentum, S. Kolodinski, P. Polakowski and J. Müller, 13th International Conference on Modern Materials and Technologies (CIMTEC), Tuscany; Italy (2014)

Acknowledgments

My deep gratitude goes to everybody who directly or indirectly supported me during my work on this thesis.

My particular gratitude goes to my doctoral advisor Prof. Dr. Thomas Mikolajick for the opportunity to prepare and complete my PhD work at NaMLab gGmbH, his patience, encouragement and helpful discussions in the course of the work.

A special thanks goes to my daily supervisor Dr. Uwe Schröder for a lot of valuable discussions, for sharing his knowledge, for patience and encouragement.

I am also grateful to Prof. Kathrin Dörr for accepting to review this thesis.

My deep gratitude goes to Dipl. Nat. Johannes Müller for laying the groundwork on the topic of ferroelectricity in HfO_2, for providing the capacitor samples with $Si:HfO_2$, for performing XRD and XPS analyses and for putting a great effort into the fabrication of the FeFET devices. Without his contribution the studies, presented in this work, would not have been possible.

My special gratitude goes to my colleges on the Ferro-team at NaMLab to Dipl-Ing. Stefan Müller, M.-Eng. Tony Schenk, Dr. Stefan Slesazeck, Dipl.-Ing. Milan Pešić and Dipl.-Ing. Karan Khullar for a lot of fruitful discussions and valuable advice, for critical reflection of my results. All these essentially contributed to my better understanding of the topic.

My deep professional and personal gratitude goes to Dr. Matthias Grube for sharing his knowledge on XRD, for proof-reading of the manuscript chapters on the ferroelectric capacitors and transistors.

Dr. Dominik Martin is thankfully acknowledged for performing PFM measurements and sharing his insights on PFM.

Dipl.-Ing. Steve Knebel is gratefully acknowledged for supporting my work in the lab.

Dr. Guntrade Roll is gratefully acknowledged for introducing me to charge-pumping and single-pulse measurements.

Dipl.-Ing. Raik Hoffman is thankfully acknowledged for supporting my measurements at Fraunhofer CNT

Dr. Thomas Gemming is thankfully acknowledged for TEM image acquisition.

My deep gratitude does to Dipl.-Phys. Steve Kupke for a lot of discussions about the origin of traps in HfO_2 and reliability of high-k metal gate stacks and for proof-reading the trapping chapter of my manuscript.

A special thanks goes to Dr. Andrew Graham for preparation of capacitor samples and especially for proof-reading of the manuscript.

The results presented within this thesis were generated within the research projects HEIKO and Cool Memory. Therefore, I would like to acknowledge the EFRE fund of the European Community and the Free State of Saxony for the financial support of these projects.

My colleges at Fraunhofer CNT and GloabalFoundries in Dresden I would like to acknowledge for their work, which made the fabrication of the $Si:HfO_2$-basierten FeFET devices possible.

I would like to gratefully thank all my colleges at NaMLab for creating a present working atmosphere, for valuable advice and assistance.

My deep personal gratitude goes to my family and friends for their emotional support in the most hopeless times and a lot of patience.

Bisher erschienene Bände der Schriftenreihe Research at NaMLab

Herausgeber: Thomas Mikolajick   ISSN 2191-7167

Alle erschienenen Bücher können unter der angegebenen ISBN direkt online
(http://www.logos-verlag.de) oder per Fax (030 - 42 85 10 92)
beim Logos Verlag Berlin bestellt werden.